Religious Violence between Christians and Jews

Religious Violence between Christians and Jews

Medieval Roots, Modern Perspectives

Edited by

Anna Sapir Abulafia

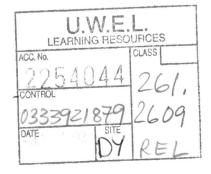

palgrave

First published 2002 by
PALGRAVE
Houndmills, Basingstoke, Hampshire RG21 6XS and
175 Fifth Avenue, New York, N. Y. 10010
Companies and representatives throughout the world

PALGRAVE is the new global academic imprint of
St. Martin's Press LLC Scholarly and Reference Division and
Palgrave Publishers Ltd (formerly Macmillan Press Ltd).

ISBN 0–333–92187–9

This book is printed on paper suitable for recycling and
made from fully managed and sustained forest sources.

A catalogue record for this book is available
from the British Library.

Library of Congress Cataloging-in-Publication Data
Religious violence between Christians and Jews : medieval roots,
modern perspectives.
      p. cm.
   Edited by Anna Sapir Abulafia
   Includes bibliographical references and index.
   ISBN 0–333–92187–9
   1. Judaism—Relations—Christianity. 2. Christianity and other
religions—Judaism. 3. Antisemitism—History. 4. Christianity
and antisemitism. 5. Judaism (Christian theology)—History of
doctrines—Middle Ages, 600–1500. 6. Violence—Religious
aspects—Christianity. I. Sapir Abulafia, Anna.

BM535 .R425 2001
261.2′6′09—dc21
                                                        2001032727

10   9   8   7   6   5   4   3   2   1
11   10   09   08   07   06   05   04   03   02

Printed and bound in Great Britain by
Antony Rowe Ltd, Chippenham, Wiltshire

Contents

PART II   ROUND TABLE DISCUSSION

List of Abbreviations

BT	Babylonian Talmud
CCCM	Corpus Christianorum Continuatio Mediaevalis
CCSL	Corpus Christianorum Series Latina
JQR	*Jewish Quarterly Review*
MGH	Monumenta Germaniae Historica
PL	J. P. Migne, Patrologiae Cursus Completus Series Latina
RHC Oc	Receuil des Historiens des Croisades, Historiens Occidentaux
RHGF	Receuil des Historiens des Gaules et de la France
RS	Rolls Series

Notes on the Contributors

Anna Sapir Abulafia is Senior Tutor and Ernst Cassel Lecturer in History at Lucy Cavendish College, Cambridge. Her field of research is twelfth- and thirteenth-century intellectual history with particular emphasis on the medieval Christian–Jewish debate. She is the co-editor (with G. R. Evans) of *The Works of Gilbert Crispin, Abbot of Westminster* (Auctores Britannici Medii Aevi, VIII [1986]) and the author of *Christians and Jews in the Twelfth-Century Renaissance* (1995) and *Christians and Jews in Dispute: Disputational Literature and the Rise of Anti-Judaism in the West* (c. *1000–1150*) (1998).

Geoffrey Alderman is Vice-President and Professor of European and Jewish History, Touro College, New York, formerly Pro-Vice Chancellor, Middlesex University, London. His current field of research is the political sociology of Anglo-Jewry. He is the author of *Modern British Jewry* (2nd edn, 1998); *London Jewry and London Politics, 1889–1986* (1989) and *The Jewish Community in British Politics* (1983).

Christopher Andrew is Professor of Modern and Contemporary History, Chairman of the Faculty of History at Cambridge University, and President of Corpus Christi College, Cambridge. He is also Chair of the British Intelligence Study Group, former Visiting Professor at Harvard, Toronto and the Australian National University, co-editor of *Intelligence and National Security*, and a regular presenter of BBC documentaries. His research in recent years has concentrated on the use and abuse of intelligence agencies in East and West. The most recent of his fourteen books are *For the President's Eyes Only: Secret Intelligence and the American Presidency from Washington to Bush* (1995) and *The Mitrokhin Archive: The KGB in Europe and the West* (1999, with Vasili Mitrokhin). The latter volume includes an analysis, based on still classified KGB files, of religious persecution in the former Soviet Union.

Deirdre Burke is Subject Co-ordinator for Religious Studies, University of Wolverhampton. Her field of research is Holocaust studies, and in particular religious and educational issues. She has published 'Attitudes to Death during the Holocaust' in *The Journal of Beliefs and Values* (1999).

Robert Chazan is currently Scheuer Professor of Hebrew and Judaic Studies at New York University. Previously, he taught at the City University of New York and the Ohio State University. His books include *Medieval Jewry in Northern France* (1974); *Church, State, and Jew in the Middle Ages* (1980); *European Jewry and the First Crusade* (1987); *Daggers of Faith* (1989); *Barcelona and Beyond* (1992); *In the Year 1096* (1996); *Medieval Stereotypes and Modern Antisemitism* (1997): *God, Humanity, and History: The Hebrew First Crusade Narratives* (2000). He is currently working on the first outpouring of medieval Jewish polemical literature, from southern France and northern Spain, during the late twelfth and early thirteenth centuries.

Jeremy Cohen is Professor of Medieval Jewish History at Tel Aviv University. Among his publications are *The Friars and the Jews: The Evolution of Medieval Anti-Judaism* (1982), *'Be fertile and increase, fill the Earth and master it': The Ancient and Medieval Career of a Biblical Text* (1989) and *Living Letters of the Law: Ideas of the Jew in Medieval Christianity* (1999).

Mark R. Cohen is Professor of Near Eastern Studies, Princeton University. His field of research is the history of the Jews in the Medieval Islamic world. Currently, he is engaged in a study of poverty and charity in the Jewish community of medieval Egypt. Among his publications are *Jewish Self-Government in Medieval Egypt* (1980); *Al-Mujtama' al-yahudi fi Misr al-islamiyya fi al-'asr al-wusta* [*Jewish Life in Medieval Egypt, 641–1382*] (1987); *The Autobiography of a Seventeenth-Century Rabbi: Leon Modena's 'Life of Judah'* (1988) and *Under Crescent and Cross: The Jews in the Middle Ages* (1994).

John D. Klier is Corob Professor of Modern Jewish History and Head of the Department of Hebrew and Jewish Studies at University College London. His research concentrates on the history of the Jews in the pre-revolutionary Russian empire. He is the author of *Russia Gathers her Jews: The Origins of the Jewish Question in Russia, 1772–1825* (1986) and *Imperial Russia's Jewish Question, 1855–1881* (1995) and (co-edited with S. Lambroza), *Pogroms: Anti-Jewish Violence in Modern Russian History* (1991). He is completing a major study of the pogroms of 1881–2 in Russia.

Gavin I. Langmuir is Professor Emeritus of History, Stanford University, California. His research interests include monotheism and anti-semitism. He is the author of *Toward a Definition of Antisemitism* (1990) and *History, Religion, and Antisemitism* (1990).

Jonathan Riley-Smith is Dixie Professor of Ecclesiastical History, University of Cambridge. His field of research is the Crusades and Latin East, particularly, at present, the Templars. His publications include *The Knights of St John in Jerusalem and Cyprus, c. 1050–1310* (1967); *The Feudal Nobility and the Kingdom of Jerusalem, 1174–1277* (1973); *What were the Crusades?* (1977; 2nd edn 1992); *The First Crusade and the Idea of Crusading* (1986) and *The First Crusaders, 1095–1131* (1997).

Israel Jacob Yuval is Professor of Medieval Jewish History at the Hebrew University, Jerusalem. He is the author of *Scholars in their Time: The Religious Leadership of German Jewry in the Middle Ages* (1988, in Hebrew). His *'Two nations in your Womb': Perceptions of Jews and Christians in the Middle Ages*, which was published in Hebrew in 2000, will be published shortly in English.

Introduction

In August 1999 the Council of Christians and Jews organized an international conference to mark the 900th anniversary of the fall of Jerusalem to the First Crusade. In stark contrast to the triumphalist ethos of the crusading armies in 1099, a number of prominent scholars came together to discuss openly with each other and with a group of religiously committed lay people the medieval roots of interfaith violence, with respect to Christians and Jews. The overall achievement of the conference was to demonstrate how important rigorous and objective historical research into violence between Christians and Jews is for creating better understanding between present-day Christianity and Judaism. Real dialogue between members of these two groups can only take place if those engaged in discussions know and respect the history and development of *both* religious/cultural traditions. This means that they must honestly confront, acknowledge and discuss the elements of both traditions which have had – and may still have – potential for engendering violence. This does not mean that anyone has to renounce what is precious to him or her. It does demand honest reflection about the impact that some traditions or beliefs might have had, or continue to have, on others. This kind of historical reflection has implications that go far beyond Christian–Jewish relations. Objective research into any kind of interfaith violence of the past forces one to engage with compelling questions concerning the potential impact of any ideology, whether or not religious in nature, which claims to grant its adherents a monopoly over truth. It urges one to think hard about the real conundrums posed by the modern concept of multiculturalism. Issues arising from ideological systems of beliefs are as pertinent to the twenty-first century as they were to the Middle Ages. They need to be addressed urgently by politicians throughout the world and by anyone in a position of leadership.

The purpose of this volume is to bring together in an accessible way the papers presented to the Conference. Readers will find a full range of approaches to the history of anti-Judaism: theological, cultural, intellectual, socio-economic, socio-psychological, etc., and they will be able to decide for themselves which approaches they find most helpful. They will find explorations of engagement by Jews with their host societies and Jewish responses to threats of violence. They will also find in-depth analyses of medieval Christian thought and spirituality. The chronological spread from medieval to modern emphasizes the importance of a medieval understanding of violence with respect to Christians and Jews, in order to achieve a modern understanding of the phenomenon. All contributions aim to be accessible to students and interested lay people. With full annotations they will, at the same time, offer new insights to specialists in the field. The bibliography offers non-specialists a selection of further secondary reading and available English translations of some of the richest source material used throughout the volume.

As each year goes by the interest in the medieval confrontation between Christians and non-Christians increases. The Christo-centric writing of medieval history is becoming rarer than it used to be. More frequently than before attention is paid to Jews, Muslims and other non-Christians. A good example of this are the second and third editions of Christopher Brooke's *Europe in the Central Middle Ages, 962–1154* (1987 and 2000), another is the approach taken in David Abulafia's volume (*c.*1198–*c.*1300) of *The New Cambridge Medieval History* (volume 5, 1999). As research in this area continues to develop, appreciation grows of how much more can be learnt about Christendom itself by investigating its attitudes towards non-Christians and what was deemed to be deviant Christian thought and behaviour. Religions and cultures have to define themselves before they can work out a detailed response to others. Nor is the study of Christian–Jewish relations simply a question of Christian attitudes towards Jews. Increasingly, scholars are studying medieval Jews and Judaism in their own right and discovering more about Jewish attitudes towards Christians.

The study of religious violence between Christians and Jews cannot confine itself to a narrow analysis of religious differences between Christians and Jews. The subject begs questions concerning the relationship between the ideas developed by the learned sectors of society and the beliefs of the less educated. It raises the issue of the nature of the interplay between the policies of those invested with religious, social and/or legal authority and the enforcement of those policies, or, to put it differently, it probes the connections between theory and practice. It involves examination into the causes of violence, the possible reasons for fluctuations in the occurrence of violence and the types of violence displayed. Most scholars would agree that with the marked exception of Visigothic Spain (in the seventh century) Jews in Latin Christendom lived relatively peacefully with their Christian neighbours for most of the early Middle Ages. Although no one would any longer label the First Crusade, which was preached in 1095 by Pope Urban II,

as the great watershed in European Jewish history, the crusade movement did bring new factors to bear on relations between Christians and
Jews. The harnessing of the concept of holy war by the papacy itself
marks an important phase in the development of western Christendom.
This development interconnected with many other important changes
in late eleventh-century European society. It is perhaps not surprising
that such significant changes started to take place in relations between
Christians and Jews in this period. In recent years scholars have done a
very great deal to find out more about all these different issues and they
have tried to relate them to the changing nature of Christian–Jewish
relations. Much work has also been produced on the significant differences between the situation in northern and in southern Europe in this
respect. All the contributors to this volume have published extensively
in different areas of this general field of enquiry. Their contributions to
this volume reflect some of their freshest views about the nature of
religious violence between Christians and Jews.

Jonathan Riley-Smith is well known for his extensive work on the
Crusades. In his 'Christian Violence and the Crusades' (Chapter 1), he
reopens the question of why persecution of Jews so often occurred in
the *preparatory* stages of the crusades which focused on Jerusalem. Why
did the preaching of crusades affect Christian actions against Jews in
the West in a way that it did not in the East? After examining the theme
of Christian holy war and the peculiar penitential nature of the crusades,
Riley-Smith posits the view that holy war has the tendency to focus
attention inwards on to the society waging war on behalf of God.
Protagonists in the exercise tend to develop the need to reform and
religiously unify their own society before engaging in battle against
external enemies. In other words, internal enemies are identified and
dealt with. Riley-Smith concludes by explaining how charismatic preachers
who were engaged in raising armies against external enemies used evoca-
tive language that encouraged violence against Jews, who were cast into
the role of internal enemies. Once crowds had been incited by religious
fervour, they were hard to control.

Robert Chazan has published prolifically on medieval Jewish history.
In his 'The Anti-Jewish Violence of 1096: Perpetrators and Dynamics'
(Chapter 2), he focuses on the First Crusade in order to identify exactly
who the perpetrators of violence against Jews were and what their
motives might have been. Through a close reading of contemporary
Hebrew material, which describes the course of the persecutions that
took place in the cities of the Rhineland in 1096, he demonstrates how
diverse the perpetrators in fact were. Some of the violence was committed

by organized crusade bands like the army of Count Emicho, other violence was perpetrated by random crusaders with or without cooperation
of the burghers; in others places burghers were the aggressors with or
without the help of crusaders; at times villagers were also involved. And
why was violence committed? Although there was no official call by the
Pope for violence against Jews, crusade sermons were rife with themes
that could spark off anti-Jewish aggression in those who had or had not
taken the cross. And once hostility had been sparked it bred further
hostility, especially where it fed off any existing latent local animosity
between Christian and Jewish burghers of a particular town. The notion
among Christians that the violence which was occurring proved con-
clusively that God had abandoned the Jews, could only encourage them
to demand of Jews they agree to be baptized or die.

Jeremy Cohen is very well known for his book *The Friars and the Jews*,
which argues that the mendicant friars played a crucial role in promot-
ing anti-Judaism from the thirteenth century onwards. In his 'Christian
Theology and Anti-Jewish Violence in the Middle Ages: Connections
and Disjunctions' (Chapter 3), he focuses on two examples of anti-
Jewish violence: the persecutions of Jews during the First and Second
Crusades and the burning of the Talmud in Paris in the 1240s. The
paper suggests some answers to the question raised by Riley-Smith's and
Chazan's papers – that is, why does religious violence take place against
Jews when Christian theology does *not* permit it? Why does harsh anti-
Jewish preaching in one period and area engender violence against Jews
and not in another? Why is it that theory and practice so often do not
follow suit? By examining letters and sermon material by St Bernard
and Peter the Venerable, and papal and royal material concerning the
trial of the Talmud in Paris, Cohen brings out the ambivalence and hesitation which the Church displayed over the question of the role of Jews
in Christian society. On the one hand, they were safeguarded by the
Augustinian principle that they served Christian society by witnessing
to the scriptural basis of Christianity; on the other, they were vilified as
embodying the evils within Christian society which needed to be
expunged. Christian ambivalence towards Jews gained an extra dimension when Christians became knowledgeable about Jewish post-biblical
writings like the Talmud. On the one hand, the Talmud was seen as
negating the biblical function Jews were supposed to play in Christian
society, and it was condemned as blaspheming Christianity and keeping Jews from Christian truth. On the other hand, sections of it were
deployed in an effort to convert Jews. Once again no unified policy was
conceived or put into practice.

My own work has up to now concentrated on the intellectual encounter between Christians and Jews. I have been particularly interested in the hardening of Christian attitudes towards Jews on account of the absorption of classical philosophical concepts into late eleventh- and twelfth-century theology. In my 'The Intellectual and Spiritual Quest for Christ and Central Medieval Persecution of Jews' (Chapter 4), I elaborate on questions brought forward in all three preceding papers by delving into the very heart of internal Christian thought and spirituality. My aim is to uncover the content of contemporary spiritual yearnings in order to pinpoint more precisely why and how central medieval Christian spirituality could lend itself to anti-Jewish thinking and behaviour. A close examination of influential Latin spiritual texts concerning Jesus Christ offers fresh insights into the development and dissemination of anti-Jewish ideas in this period.

Israel Yuval's work has focused on medieval Jewish responses to Christianity. In his ' "They tell lies: you ate the man": Jewish Reactions to Ritual Murder Accusations' (Chapter 5), he delves into Hebrew literary material in order to gauge how Jews reacted to Christian ideas about Christ and the Eucharist and Christian accusations against Jews. Using Latin and Hebrew sources, Yuval discusses the development of the ritual murder accusation in, in particular, France and Germany. He argues that the distinction between the accusation of ritual murder (in which Jews are accused of crucifying a Christian child) and the blood libel (which involves the accusation of ritual cannibalism) is not as clear-cut as scholars have supposed. Yuval demonstrates that the responses by medieval Jews to these accusations tended to internalize the motifs of their accusers in order to emphasize the sanctity of Jewish martyrdom at the hands of Christians. Throughout his article Yuval highlights remarkable instances of common usage of symbolic language by Christians and Jews.

Mark Cohen has explored in depth the differences between Jewish life in medieval Christendom and Islam. His book *Under Crescent and Cross*, which argued that in broad terms medieval Islam provided a preferable host society for Jews than medieval Christendom, provoked varied responses. In his 'Anti-Jewish Violence and the Place of the Jews in Christendom and in Islam: a Paradigm' (Chapter 6), he takes the opportunity to respond to those who took issue with his findings by looking more closely at the different environments Jews experienced within medieval Christendom. Mark Cohen starts by creating a paradigm on the basis of a broad comparison between medieval Christian–Jewish and medieval Christian–Muslim relations. He highlights the marked difference between Christian and Muslim attitudes towards Jews. His paradigm asserts that levels of anti-Jewish violence are connected to 'the totalitarianism of religious exclusivity', combined with mitigating or exacerbating economic, legal and social circumstances. He then applies this paradigm to early medieval northern Europe, southern France and Italy, Reconquista Spain, Byzantium, medieval Poland and late medieval Islam in order to explain why some periods and places display less harsh anti-Judaism than others. Until well into the thirteenth century the Jews of Sefarad (Spain and neighbouring lands) were very much more integrated into their host societies than in Ashkenaz (Germany and neighbouring lands).

In contrast to the preceding papers, Gavin Langmuir's 'At the Frontiers of Faith' (Chapter 7) engages with the nature of violence against Jews by looking beyond what is seen by him as an overlay of religious issues. Langmuir is well known for his two books *Toward a Definition of Antisemitism* and *History, Religion and Antisemitism*, in which he discusses antisemitism as the irrational version of anti-Judaism. Langmuir defines religious violence as 'the exertion of physical force on human beings, whether by societies or individuals, that is primarily motivated, and explicitly justified, by the established beliefs of their religion'. Religious violence against Jews was motivated by the fact that their God is not the Trinity of Christians, and Jews suffered religious violence at the hands of Christians during and after the First Crusade. But according to Langmuir, unambiguously religious violence started declining in the twelfth and thirteenth centuries to make way for a different kind of violence. Although this violence was also couched in religious terms, it was fuelled by completely irrational accusations like ritual murder, blood libel and host desecration. The roots of this kind of violence, according to Langmuir, are not, properly speaking, religious. Langmuir calls this violence psychopathological and he asserts that it was 'motivated and explicitly justified by the irrational fantasies of paranoid people whose internal frontiers of faith were threatened by doubts they did not admit'. Religious language was used, but for Langmuir it did not really concern religion at all.

Moving eastward from Mark Cohen's discussion of the pluralism of medieval Poland and forward in time, John Klier contributes his expertise on the Jews of Russia to this volume. In his 'Christians and Jews and the "Dialogue of Violence" in Late Imperial Russia' (Chapter 8), he makes extensive use of the archival material which has recently become available to unravel the true facts of the pogroms against Jews of 1881–2 in the Russian Empire. Counter to accepted views, Klier argues provocatively that these pogroms were not organized by the state and were not in the first place motivated by religious differences. The paper argues that the pogroms can only be properly understood when they are studied in the full framework of the violent nature of Russia's multi-ethnic and socially diverse society. In the vein of many of the medieval papers, Klier looks carefully at the broader context of Jewish life in Russia by examining the social and economic position of the Jews in the Pale of Settlement. And he emphasizes the self-defence Jews were able to muster during these riots.

It is equally important to make plain what the papers of the Conference did not cover. The Conference could not and, indeed, did not intend to cover the whole spectrum of violence between Christians and Jews, including in-depth analyses of early modern and modern forms of antisemitism, for which questions of race and nationalism are of obvious importance. The chosen topic was religious violence and as such the emphasis of the Conference was on the medieval period. But that does not mean to say that religion ceased to play a role after *c.* 1500 and that medieval forms of violence against Jews have nothing to teach students of later guises of anti-Jewish violence. However important factors like nationalism and racism became within particular social, political and economic contexts, religion still remained significant if only to provide a reservoir of vocabulary with which to articulate hatred and facilitate violence. For modern historians it is also important to be reminded that medieval violence against Jews was not just religious either. Religion can never be a fixed category; it is an institutional expression of a set of beliefs, on the one hand, and the internalisation of that expression among those who are members of that institution, on the other. Both the expression and internalization of beliefs and the interaction between religious institutions and their members will vary according to prevailing social, economic and political circumstances. Mark Cohen's paper is an excellent example of how important it is to study this kind of interplay of factors in order to understand the different levels of intensity of medieval violence against Jews in time and place. Gavin Langmuir's paper discusses at length the different forms Christianity took in the medieval period. He also illustrates how the vocabulary of religion can overlay other motives for violence. John Klier picks up this strand in a modern setting. However important religious language is, it is essential for all periods to investigate what other factors led to any particular instance of religious violence.

All these points were fully discussed in the Round Table which was part of the conference and was chaired by Professor Christopher

Andrew. Geoffrey Alderman, in Chapter 10, highlights the religious motifs of twentieth-century anti-Jewish violence in Britain ('Some Thoughts on Anti-Jewish Violence in Modern Britain') Christopher Andrew's contribution, 'Religions Violence in Past and Future Perspective' (Chapter 9), makes plain how foolish it would be to ignore the strength of religious convictions in modern forms of violence, whatever the racial, socio-economic or political categories upon which they are grafted. The discussion made it plain how important it is for modern historians to have a better understanding of the many different guises of medieval anti-Judaism. Equally important is the need to study carefully the Jewish input into the history of religious violence. Seeing Jews as the perennial victims without any control over their own destinies is as unhelpful as it is untrue. Jonathan Riley-Smith, in 'Religious Violence' (Chapter 11), demonstrates how knowledge of the medieval meaning of holy war is invaluable in analysing its modern counterparts. The overwhelming conclusion of the discussion was that anyone studying religious violence with respect to Christians and Jews could not just investigate objectively what happened long ago. Part and parcel of their work is to draw lessons from it in order to help present-day society come to terms with whatever its tradition of violence is and learn how to contain it. As Deirdre Burke makes plain in 'Religious Violence: Educational Perspectives' (Chapter 12), the role of education is vital.

The Council of Christians and Jews states that it 'brings together the Christian and Jewish Communities in a common effort to fight the evils of prejudice, intolerance and discrimination between people of different religions, races and colours, and to work for the betterment of human relations, based on mutual respect, understanding and goodwill'. It is hoped that the collection of these papers will stimulate informed and constructive discussions about the nature, causes and effects of religious violence with respect to Christians and Jews. If the volume leads to fresh insights which prove to be useful not just in an academic sense, but on a practical level too, it will have served its purpose well.

Cambridge   A.S.A.

Part I

Studies on Religious Violence between Christians and Jews

the emperor Frederick I and the German bishops,[5] and in 1189 and 1190 there were riots and killings throughout England, the most notorious being in York.[6] There was renewed persecution in western France in 1236, again marked by attempts to convert by force, at a time when another crusade was being preached.[7] As late as 1320, the masses on the second Crusade of the Shepherds claimed to be exacting vengeance for the crucifixion and were forcing baptism on Jews over a large part of southern France.[8] I am not suggesting that the immediate causes of these events were necessarily the same, but the preaching of a crusade on behalf of Jerusalem was one element common to all of them, as were the themes of vengeance and conversion.

But two things puzzle me. First, it is hard to reconcile calls for vengeance with demands for conversion: one is an act of retaliation, the other would have been regarded by the perpetrators as the conferment of a benefit. Secondly, emotional anti-Judaism seems to have featured in a European context, but not, or not so much, in a west Asian one. This requires a moment's explanation, since it is often supposed that Jews were targeted throughout the course of each crusade. Over the last 30 years, owing to the work of Professor S. D. Goitein, the evidence for a massacre of Jews in Jerusalem in 1099 has largely evaporated;[9] even in relation to the Muslims the estimates of the dead have now fallen drastically.[10] It is true that the earliest crusaders seem to have pursued a policy – not unlike what we now call ethnic cleansing – by which non-Christians were driven, by terror if need be, from places considered to be of religious or strategic significance,[11] but this policy did not distinguish the Jews from others, and specific anti-Judaism seems to have vaporized – or at least did not find expression – once the armies had left Europe. The fact that focused persecution occurred in the preparations for, rather than in the course of, crusades adds to our difficulties. It is, for example, less easy to explain why the *locus* of Jerusalem was the trigger it clearly was. The first serious outbreak of violence against Jewish communities in northern Europe in the central Middle Ages, which had occurred 85 years before the First Crusade, seems to have been set off by the vandalization of the Holy Sepulchre in Jerusalem on the orders of the Fatimid caliph al-Ḥākim.[12] When the news of the fall of Jerusalem to Saladin reached the west in 1187, it was reported that the reaction of Christians in Germany was to say: 'The day for which we have waited has arrived – the day for killing all Jews'.[13] But while persecutions associated with crusading seem to have been linked to crusades to the east, rather than to those against heretics or political opponents of the papacy or pagans in the Baltic region or the

Mongol penetration into eastern Europe, how is one to explain the fact that, at any rate after 1110, there was relative toleration in crusader Palestine?[14]

I am going to suggest an alternative explanation for the persecutions. After outlining the traditions of Christian holy war and crusading in particular, I will dwell on the tendency of holy war to turn inwards and the appeal introspective violence appears to have had for lay men and women; and on the failure of the church leaders to control public emotions, a failure which can to some extent be explained by deficiencies in the process of crusade recruitment.

Christian holy war

Crusade theory was an adaptation of a much more ancient tradition of Christian sacred violence, the foundations of which had been laid by theologians in the fourth and fifth centuries, in a Roman empire which was now Christian, with a government that had assumed the responsibilities of its pagan predecessor for assuring internal order and defence. For these theologians, violence which they believed had been commanded at times in the past by God could not be intrinsically evil. It had to be morally neutral, ethical colouring being attached to it by the intentions, bad or good, of the perpetrator. The rightness of the intention that lent violence its moral standing had to be measured in terms of love, and those who authorized and took part in it had, therefore, to be careful to employ only such force as was necessary. Augustine of Hippo, the greatest of the early theoreticians, returned to the theme of loving violence time and time again, for it was this that provided the basis for his justification of the physical suppression of heresy. It was right, and a sign of love and mercy in imitation of Christ, for a loving church in collaboration with a loving state to force heretics from the path of error for their own benefit, compelling them to goodness in the same way as in Christ's parable the host at the feast had sent out his servant to force those on the highways to come to the banquet.

Violence could accord with the divine plan and be pleasing to God when employed as a means of achieving justice. It required a just cause, of course, which meant that it could only be used in just reaction to intolerable injury, usually taking the forms of aggression or oppression. Its reactive nature meant that in theory, although by no means always in practice, the initiative had to lie with the aggressor. Missionary wars, for example, were not theoretically permissible.

All rulers, even pagans, were divine ministers, but the Christian Roman emperors were believed to be the special representatives of God, who had put the temporal power of the empire at the Church's disposal for its defence. God could, however, bypass his ministers on earth by ordering the use of violence personally and directly. The precedents cited in support of a divinely ordained violence came mostly from the Old Testament, but ambivalence on this issue was also to be found in the New Testament. A defining feature of Christian sacred violence, therefore, was the fact that it was perpetrated on God's indirect or direct authority. The agent involved in such violence was believed to be performing a service to God, an idea which appeared early in Christian thought in the Epistle to the Romans, in which St Paul justified the coercive sanctions at the disposal of the Roman state.

> For [the power] is the minister of God to thee for good. But if thou do that which is evil, be afraid; for he beareth not the sword in vain: for he is the minister of God, a revenger to execute wrath upon him that doeth evil.   (Romans 13: 4)[15]

Crusading

I have outlined patristic thought not so much as it was, but as it appeared once eleventh- and twelfth-century writers had tidied it up, removing the contradictions and giving it a coherence it had not had before. Crusade propagandists took trouble to conform their arguments to the patristic criteria of right intention, just cause and legitimate authority,[16] because throughout the history of the crusading movement people had to be persuaded that the danger to which they were going to expose themselves was worthwhile. Sometimes the arguments could be specious, as in Pope Innocent III's statement that crusading in Livonia was legitimate because 'the church' there – a handful of converts – was coming under threat from indigenous paganism[17] but the fact that justification was needed shows how seriously the issues were being treated. Of course, there must have been many who felt that the mere facts of Islam, shamanism or paganism provided ample justification, but enough people needed to be convinced by argument for the just cause to feature at length in almost every crusade proclamation.

The crusade was penitential and was at first associated with pilgrimage to Jerusalem, the most penitential goal of all and a place where devout Christians went to die, which may be why so many crusaders were old men. The cross was invariably enjoined on them as a penance. They were not supposed to travel gloriously to war, but to dress simply as pilgrims, with their arms and armour carried in sacks on pack-animals. In 1099, with their goal achieved, most of the survivors of the First Crusade apparently disposed of their arms and armour and returned to Europe carrying only the palm fronds they had collected as evidence that they had completed their pilgrimage.[18] The idea of penitential war was unprecedented in Christian thought, as conservative opponents of the papacy pointed out at the time. It gave combat an entirely new dimension and it is not surprising that contemporaries came to view the taking of the cross as in some sense an alternative to entry into the religious life; indeed, comparisons between monasticism and crusading were being made even before the first armies marched[19] and it was a natural progression that led, within a quarter of a century, to professed religious in the military orders themselves engaging in war.[20]

Crusade propagandists expatiated on the idea of war at Christ's command mediated by the pope as Christ's vicar on earth. From the start the crusading armies were 'armies of Christ' or 'of God' and the crusaders were 'knights of Christ'. The achievements of the First Crusade, in which an army of knights without horses and pack-animals – they had nearly all died within a year – without any overall commander or system of provisioning and encumbered by non-combatants, marched into Asia and after two terrible years took Jerusalem, two thousand miles from home, struck those who took part and those who stayed in the west as so miraculous that they provided evidence for divine intervention.[21] Once this conviction, reinforced by the reports of the appearances to the crusaders of signs in the heavens and of visions of Christ, Our Lady, the saints and the ghosts of their own dead, was fixed in the minds of western men and women, nothing, not even catastrophic failure in the future, could expunge it, because failure could always be explained not as a demonstration that crusading was against God's wishes, but by the fact that the instruments of that particular divine plan were too unworthy to carry it out.

It should be stressed that nothing in the traditions of Christian violence or crusade theory could have justified theoretically, or would have led inevitably to, the persecution of Jewish communities, provided that no acts of those communities could be described as being presently injurious. Responsible men agreed that they were not, although the wild fantasies about them that circulated during the riots could have been concocted, perhaps unconsciously, to provide a just cause for the use of force against them.[22]

The internalization of holy war

Sacred violence can take several forms, of course, but the category of holy war to which crusading belongs – extraliminal in the sense that it is initially proclaimed against an external force – seems to have a tendency, whatever the religion involved, to turn inwards sooner or later and to be directed against the members of the very society which has generated it. In Islam there developed a tradition which distinguished the greater *jihad* (holy struggle), involving ethical and spiritual renewal, from the lesser, which related to war. In the twelfth-century Near East, where a *jihad* was being waged against the crusaders, there was a movement for moral rearmament, which came to prominence under Nur ad-Din, the Muslim ruler of Damascus and Egypt, and was marked by the building of *madrasas* (religious colleges) and the suppression of heresies, particularly Shi'ism.[23] The climax of this movement came around 1300 with the remarkable and charismatic Ibn Taymiyya, for whom the priority of the *jihad* was at that stage not to wage war in the *dar al-Harb* (land of war), but to purge the Sunni world of infidels and heretics. He condemned the Shi'ites as interior enemies 'even more dangerous than the Jews and the Christians' and he wanted holy war to be waged pitilessly against them. For him the *jihad* was a force which at the same time would renew individual spirituality and create a society dedicated to God which could then triumph over the world.[24] In nineteenth-century Java, Dipanagra, the leader of a religious revolt against all foreigners and unbelievers which after a short purifying period of violence ought to have culminated in his rule as the millenarian 'just king', wanted to impose strict Islamic uniformity.[25] In twentieth-century Vietnam, the Hoa Hao sect, which had inherited radical millenarian warlike Buddhism and flourished among the peasants in the Mekong Delta, believed that individual salvation could only occur in the context of collective salvation; religion and politics were fused and were marked by extreme xenophobia.[26] A historian who has studied various millenarian anti-colonial movements – in Java, New Zealand, India, German East Africa and Burma – has concluded that in all of them there was the desire to purge alien groups and the conviction that the colonial overlords, their assistants and all who refused to follow the prophet concerned would perish or be driven away.[27]

Apocalyptic ideas comprised only an undercurrent in crusading, except perhaps for a short period at its start and in the thought of a minority of individuals, such as Pope Innocent III,[28] but in other respects the movement shared many of the features I just described. The first sign of its becoming officially introspective came in 1135 when Pope Innocent II presided over a council at Pisa which decreed that those who fought the pope's enemies in the west – in this case the south Italian Normans and the anti-pope Anacletus – 'for the liberation of the church' should enjoy the crusade remission of sins.[29] Some time later Peter the Venerable, the influential abbot of Cluny, was prepared to argue that violence against fellow Christians could be even more justifiable than the use of force against infidels.

> Whom is it better for you and yours to fight, the pagan who does not know God or the Christian who, confessing him in words, battles against him in deeds? Whom is it better to proceed against, the man who is ignorant and blasphemous or the man who knows the truth and is aggressive?[30]

The belief that any chance of extraliminal victory could be vitiated by corruption or divisions at home, so that only when society was undefiled and was practising uniformly true religion could a war on its behalf be successful, was being widely expressed following the disasters, including the loss of Jerusalem, which overtook the Christian settlements in Palestine in 1187.

> It is incumbent upon all of us [wrote Pope Gregory VIII] to consider and to choose to amend our sins by voluntary chastizement and to turn to the Lord our God with penance and works of piety; and we should first amend in ourselves what we have done wrong and then turn our attention to the treachery and malice of the enemy.[31]

From the Fourth Lateran Council in the early thirteenth century to the Council of Trent in the middle of the sixteenth, every general council of the Church was officially summoned on the grounds that no crusade could be really successful without a reform of the Church and of Christendom.

Everywhere one looks in the west around 1200 one can find evidence of a drive to impose uniformity on a society which was already remarkably monocultural, and it is no coincidence that introspective crusades came to the fore. When calling for a crusade against the heretical Cathars in 1208 – in propaganda replete with imagery of uncleanness and disease – Pope Innocent III summoned

> knights of Christ . . . to wipe out the treachery of heresy and its followers by attacking the heretics with a strong hand and an outstretched

arm, that much more confidently than you would attack the Muslims because they [the heretics] are worse than them.[32]

These crusades were often preached in the name of extraliminal war against Islam. Innocent accused Markward of Anweiler, against whom he proclaimed a crusade in 1199, of threatening his preparations for a campaign in the east.

We concede to all who fight the violence of Markward and his army the same remission of sins which we grant to all who arm themselves to fight the perfidy of the Muslims in defence of the eastern province, since aid to the Holy Land is hindered through his actions.[33]

Although this was obviously not the first occasion in which force had been authorized against heretics or used against the pope's political enemies – it was not even the first time war had been waged on them – the redirection against enemies within Christendom of armies of penitents originally engaged to confront external threats was a novelty, particularly since men who had taken the cross for the east found themselves being pressurized to commute their vows in favour of internal police actions.[34]

So the Christian crusade, like the Islamic *jihad*, could lead to strivings for religious uniformity. One's first reaction is that this must have created a dangerous environment for Jewish communities in western Europe and it is certainly the case that Jewish leaders became very nervous when crusades were in preparation. They were right to be apprehensive, although the irony is that they were members of the only alien community within society which was officially protected. In Islam the law of *dhimma* (covenant under which protected people lived) was a kind of bulwark for them,[35] while in Christendom the concept of the just cause – that violence could only be reactive – and the belief that it was part of the divine plan that Jews should survive in a servile condition as providential witnesses made it impossible for church leaders to tolerate the use of force against them. Ibn Taymiyya, who hated Jews and Christians, argued that while the goal of Muslims must be the eventual eradication of their imperfect monotheism, this should be achieved not by persecution, but by the strict application of the *dhimma* regulations.[36] His attitude echoed to some extent that of Peter the Venerable, who in the course of a diatribe against Jews in a letter to the king of France in 1146, wrote that in spite of everything God did not want Jews killed; on the other hand, they should be punished for their wickedness in a suitable manner, by having their profits confiscated; and the proceeds should be used to help finance the crusade being prepared by the king.[37]

The writings of church leaders have often been analysed. Here I need only say that the message is consistently the same. The Jews constitute an alien group; they live 'in our midst', but in many ways they are worse than Muslims. Although the blood of Christ cries out against them, they are suffering for what their ancestors did, their very existence being testimony to the truths of Christianity, and because there must be the expectation of their eventual, peaceful conversion.[38] One is faced by a curious situation in which leading churchmen were consistently expressing their abhorrence of Judaism, often in emotive terms, while at the same time 'ring-fencing' its adherents. They regarded the Jews as deviants from the truth, but, unlike other deviants such as heretics, rather ungraciously forbade steps being taken against them.

They were forced into this situation, which would have been ridiculous if the consequences had not been so tragic, because the pressure for introspective violence being extended to all alien communities was coming from the faithful themselves. I used to think that the forced baptisms which so horrified the Rhineland Jews in 1096 were manifestations of millenarianism, because otherwise it was hard to see how they could be reconciled with the calls for vengeance. There were millenarian ideas around at the time: a Hebrew source reported what seems to have been a garbled account of the ambition of the Jews' arch-tormenter, Emich of Flonheim, to become the last emperor of Jerusalem[39] and Guibert of Nogent put an invented speech into the mouth of Pope Urban II in which the conquest of Jerusalem was justified as a necessary step on the road to the Last Days.[40] The conversion of the Jews to Christianity could well have been another element, since it was present in what was admittedly a rather confused apocalyptic tradition. But I am now not so sure, not least because millenarianism did not feature strongly in later crusading, as I have already argued. I am now inclined to think that, in the context of the calls to cleanse and purify the Christian sanctuaries in Palestine which permeate the sources,[41] baptisms were being forced on the Jews with the aim of creating a uniformly Christian society by eliminating their religion. The writer Ekkehard of Aura reported that the intention of the persecutors was to wipe out the Jewish communities or compel them to enter the Church,[42] and there are other contemporary references to the abomination of allowing alien forces to coexist with Christians within western Europe.

It was unjust for those who took up arms against rebels against Christ to allow enemies of Christ to live in their own land.[43]

We wish to attack the enemies of God in the east, once we have crossed great tracts of territory, when before our eyes are the Jews, more hostile to God than any other race. The enterprise is absurd.[44]

It used to be believed that the expression of such extreme opinions, devoid of the qualifications put on them by leading churchmen, were characteristic of the masses. But nobles seem to have been directing the massacres of 1096, even if they were in a minority,[45] and throughout the central Middle Ages it is not hard to find similar views being held by individuals who ought to have known, or could have been advised, better. Many, perhaps most, of the *crucesignati* (wearers of the cross) who rioted in Rouen in the winter of 1095–6[46] and in Mainz in March 1188[47] were knights. The story in Matthew Paris's *Chronica maiora* (*Great Chronicle*) that King Louis IX's decision to expel Jewish usurers from his kingdom, transmitted to France in 1253 while he was still in Palestine, was an emotional reaction to the mockery of his Muslim gaolers, who had asked him why as a Christian he tolerated Christ's killers in his land,[48] should not be dismissed out of hand, given other rather simple opinions we know the king held. He believed, for example, that his brother Robert of Artois had died a martyr's death in the battle of Mansurah in 1250[49] and there is evidence for the embarrassment felt about this by the senior churchman on the crusade, the papal legate Odo of Châteauroux.[50] At any rate, it is indicative that Louis's action against Jews should have been associated with the failure of his crusade. One cannot avoid concluding that the call to crusade sparked a range of emotional responses which transcended class.

A failure of leadership

So far, I have suggested that while there was nothing in the theoretical traditions of holy war or the crusade which would have led inevitably to the persecution of Jews, the tendency for such a movement to turn inwards and seek to purify its own ground may have provided a motive for the atrocities; it is an indication of the close association between introspective and extraliminal violence that, in a role reversal, the masses on the Children's Crusade of 1212, whose goal was Jerusalem, seem to have been inspired by the preaching for the Albigensian Crusade against heretics in southern France.[51] It remains for me to try to explain why the leaders of Christianity proved to be so incapable of preventing such outbreaks.

They found themselves in an exceptionally weak position. It is hard to know how conscious they were of an older tradition, in which bishops had occasionally expelled from their cathedral towns Jews who had refused baptism;[52] public knowledge of this could have undermined their present stand. More importantly, it was not helpful for them on the one hand to use the same language as the general public – that Jews were alien blasphemers, inveterate enemies, guilty of the blood of Christ – but, on the other hand, to conclude by demanding that nevertheless they should be spared, for opaque reasons which ordinary laymen must have found hard to understand.

The nature of crusading, moreover, made their position impossible. A man or woman became a crusader by taking a vow,[53] which in canon law had to be a voluntary act, committing the individual concerned to make a pilgrimage, a penitential exercise which had to be open to all, to the sick and feeble as well as the healthy and strong, to sinners as well as to saints; indeed especially to sinners. There could therefore be no watertight system for screening volunteers, who might well include psychopaths. In the thirteenth century the unsuitable could sometimes be persuaded to redeem their vows; and the absence of any system of selection did less damage than one might have supposed, because the magnates, at least, did not want undisciplined men in their households. The fact, however, that the secular leadership of any crusade consisted entirely of volunteers made adequate chains of command hard to establish, which is one of the reasons why so many expeditions spun off course or ended up achieving very little. Most crusades, unless they were relatively small, under acknowledged leaders, were run by committees made up of the great lords together with a papal legate. It was hard to get these often proud and touchy men to agree on any course of action, partly because they themselves could never make decisions independently of their own subordinates who were, like them, volunteers and who, unless they were associated with the greater lords by ties of family or clientage back home, were serving in their contingents out of choice. The lesser nobles and knights could easily transfer their loyalties to another lord or even abandon the crusade altogether if they thought they were not being properly led. All early crusades were characterized by indiscipline and by a kaleidoscopic shifting of allegiances as minor lords moved from one contingent to another, or armies and individuals came and went.[54]

If indiscipline was a feature of the armies on the march, the process of recruitment, which provided the context of the persecutions, must have been anarchic. Near hysteria was being generated deliberately by crusade preachers, who used every technique they could muster to spark spontaneous reactions from their audiences. On the first recruiting drive of all, Pope Urban II preached in French country towns, surrounded by a flock of cardinals, archbishops and bishops, whose riding households must have been immense, after processing through the streets crowned with his tiara. His most famous sermon, his proclamation of war, was made out of doors, in a field outside Clermont, and at the end of November.[55] Bernard of Clairvaux also preached the cross in the open air 50 years later.[56] In the 1190s preachers stood before a huge canvas screen on which were painted Muslims on horseback desecrating the Holy Sepulchre in Jerusalem.[57]

The preachers often chose not only a theatrical environment, but also a significant date in the calendar. On his journey round France Pope Urban timed his arrival in towns to coincide with great patronal feasts: he was at St Gilles for the feast of St Giles, at Le Puy, the greatest Marian shrine of the time, for the feast of the Assumption and at Poitiers for the feast of St Hilary.[58] In 1188, for his most important sermon in Germany, the papal legate Henry of Marcy chose the fourth Sunday in Lent, Laetare Sunday, the introit of which begins, 'Rejoice Jerusalem and come together all you that love her. Rejoice with joy you who have been in sorrow'; it was followed by the anti-Jewish riot which had to be suppressed.[59] In 1291 the archbishop of York, employing Dominicans and Franciscans from 13 communities, organized preaching rallies in 37 places in his diocese, to be held simultaneously on 14 September, the Feast of the Exaltation of the Cross.[60]

Proceedings would begin with Mass being sung in the presence of as many senior ecclesiastics from the region as could be collected together.[61] Then any papal general letter in which Christians were summoned to a particular crusade would probably be read and translated: it was usual for crusade preachers to carry copies of it in their saddle-bags. This explains the highly emotional words with which so many of these letters opened.

> On hearing with what severe and terrible judgement the land of Jerusalem has been smitten by the divine hand, we and our brothers have been confounded by such great horror and affected by such great sorrow that we could not easily decide what to do or say; over this situation the psalmist laments and says: 'Oh God, the heathens are come into thy inheritance'.
>
> (Pope Gregory VIII, October–November 1187)[62]

Because at this time there is a more compelling urgency than there has ever been before to help the Holy Land in her great need and because we hope that the aid sent to her will be greater than that which has ever reached her before, listen when, again taking up the old cry, we cry to you. We cry on behalf of him who when dying *cried with a loud voice* on the cross, *becoming obedient* to God the father *unto the death of the cross*, crying out so that he might snatch us from the crucifixion of eternal death . . . He has granted [men] an opportunity to win salvation, nay more, a means of salvation, so that those who fight faithfully for him will be crowned in happiness by him, but those who refuse to pay him the servant's service they owe him in a crisis of such great urgency will justly deserve to suffer a sentence of damnation on the Last Day of severe judgement.

(Pope Innocent III, 19–29 April 1213)[63]

The preacher would then launch into his homily.[64] It was common for this to be based at least partly on the letter which had just been read. In his *De predicatione sancte crucis*, a portable handbook for preachers of the cross written in the late 1260s, Humbert of Romans, arguably the greatest thirteenth-century master general of the Order of Preachers after Dominic, suggested that each sermon should end with an *invitatio* (incitement), marked in the margin of his manual, in which the preacher was to implore his listeners to take the cross. He gave 29 examples,[65] but *invitationes* must usually have been extempore and we can get some idea of how passionate they could become from the report of a sermon preached in Basel by Abbot Martin of Pairis on 3 May 1200.

And so, strong warriors, run to Christ's aid today, enlist in the knight-hood of Christ, hasten to band yourselves together in companies sure of success. It is to you today that I commit Christ's cause, it is into your hands that I give over, so to speak, Christ himself, so that you may strive to restore him to his inheritance, from which he has been cruelly expelled.[66]

Humbert's *invitationes* each end with the word *cantus* and he explained in his introduction that an *invitatio* should be accompanied by a hymn.[67] The hymns must have been sung as men came forward to commit themselves. As each man made his vow he was presented with a cloth cross and he was supposed to have it attached to his clothes at once.[68] This aspect of the proceedings needed careful preparation, because otherwise there would have been confusion – at Vézelay in

1146 so great was the enthusiasm that the stock of made-up crosses ran out and the preacher, Bernard of Clairvaux, had to tear his habit into strips to provide additional ones[69] – but crusaders were expected to go on wearing the crosses they had been induced to take until they came home with their vows fulfilled. Pope Urban seems to have intended them to be distinctive and it is easy to forget how visible they must have been.[70]

In this emotional environment, preachers, trying to meet the aspirations of the society they were addressing, liked to describe Christ in anthropomorphic terms. In the early decades he was commonly portrayed as the father of a noble household who had lost his patrimony and was calling on his sons to help him recover it.

> I address fathers and sons and brothers and nephews. If an outsider were to strike one of your relations down would you not avenge your blood-relative? How much more ought you to avenge your God, your father, your brother, whom you see reproached, banished from his estates, crucified![71]

Christ was also described as a lord demanding military service from his subjects, an image which was especially popular around 1200, when the pope of the time was making great play of feudal themes and preachers were developing in their sermons the language employed in his encyclicals.

> The Lord really has been afflicted by the loss of his patrimony. He wishes to test his friends and to see whether his vassals are faithful. If anyone holds a fief of a liege-lord and deserts him when he is attacked and loses his inheritance, that vassal should rightly be deprived of his fief. You hold your body, your soul, and everything you have from the highest emperor. Today he has had you summoned to hurry to his aid in battle.[72]

This personification of Christ could lead crusaders to believe that they were being summoned on his behalf to a blood-feud or a feudal war, with the results I have already described.

The tradition of theatrical crusade preaching lasted throughout the Middle Ages and beyond. The techniques which persuaded audiences to commit themselves to a strenuous, expensive and dangerous activity also so heightened emotions that forces were unleashed which could not be controlled. Two levels coexisted in the crusading movement, one institutional and the other charismatic. The institution needed charisma if a crusade was to be launched at all, because no campaign could be fought without at least a nucleus of volunteers. So the preachers were to some extent in charisma's thrall and charisma could demand recourse to decontaminating introspective violence.

Notes

1. R. Chazan, *European Jewry and the First Crusade* (Berkeley and Los Angeles, CA, 1987) pp. 62–3.
2. Mainz Anonymous, 'The Narrative of the Old Persecutions', in *The Jews and the Crusaders: The Hebrew Chronicles of the First and Second Crusades*, ed. and trans. S. Eidelberg (Madison, WI, 1977) p. 99.
3. Solomon bar Simson, 'Chronicle', in *The Jews and the Crusaders*, ed. Eidelberg, p. 25.
4. Ephraim of Bonn, 'Sefer Zekhirah or The Book of Remembrance', in *The Jews and the Crusaders*, ed. Eidelberg, pp. 121–2. For Bernard of Clairvaux, see most recently R. Chazan, *Medieval Stereotypes and Modern Antisemitism* (Berkeley and Los Angeles, CA, 1997) pp. 42–6.
5. R. Chazan, 'Emperor Frederick I, the Third Crusade and the Jews', *Viator*, **8** (1977) 83–93.
6. See R. B. Dobson, *The Jews of Medieval York and the Massacre of March 1190*, Borthwick Papers, no. 45 (York, 1974).
7. S. Simonsohn, *The Apostolic See and the Jews: Documents, 492–1404*, Pontifical Institute of Mediaeval Studies: Studies and Texts, 94 (Toronto, 1988) pp. 163–4.
8. M. Barber, 'The Pastoureaux of 1320', *Journal of Ecclesiastical History*, **32** (1981) 143–66.
9. See, especially, S. D. Goitein, 'Geniza Sources for the Crusader Period: a Survey', in B. Z. Kedar, H. E. Mayer and R. C. Smail (eds), *Outremer: Studies in the History of the Crusading Kingdom of Jerusalem Presented to Joshua Prawer* (Jerusalem, 1982) pp. 306–14.
10. This will be demonstrated in a forthcoming publication by Professor B. Z. Kedar.
11. J. S. C. Riley-Smith, 'The Latin Clergy and the Settlement in Palestine and Syria, 1098–1100', *Catholic Historical Review*, **74** (1988) 546–9.
12. See D. F. Callahan, 'Ademar of Chabannes, Millennial Fears and the Development of Western Anti-Judaism', *Journal of Ecclesiastical History*, **46** (1995) 23–4; J. S. C. Riley-Smith, *The First Crusaders* (Cambridge, 1997) pp. 25–6.
13. Chazan, 'Emperor Frederick I', p. 87.
14. See, especially, J. Prawer, *The History of the Jews in the Latin Kingdom of Jerusalem* (Oxford, 1988) *passim*.
15. See J. S. C. Riley-Smith, 'Crusading as an Act of Love', *History*, **65** (1980) 185–7; F. H. Russell, *The Just War in the Middle Ages* (Cambridge, 1975) pp. 16–26.

16. See, for example, Humbert of Romans, *De predicatione sancte crucis* (Nuremberg, 1495) Chapters 2 and 45.
17. Innocent III, 'Opera Omnia', PL 214 coll. 739–40.
18. Albert of Aachen, 'Historia Hierosolymitana', RHC Oc 4 pp. 499–500.
19. For the penitential nature of crusading, see Riley-Smith, *The First Crusaders*, pp. 53–75; J. S. C. Riley-Smith, *The First Crusade and the Idea of Crusading* (London, 1986) pp. 135–52.
20. See Thomas Aquinas, *Summa Theologiae*, 2a2ae, qu. 188, art. 3, ed. and trans., Blackfriars edn, vol. 47 (London, 1973) pp. 192–3. For the revised date for the foundation of the Templars, see R. Hiestand, 'Kardinalbischof Matthäus von Albano, das Konzil von Troyes und die Entstehung des Templerordens', *Zeitschrift für Kirchengeschichte*, **99** (1988) 295–325.
21. Riley-Smith, *The First Crusade and the Idea of Crusading*, pp. 139–43.
22. See, for example, Mainz Anonymous, 'The Narrative', p. 102; Solomon bar Simson, 'Chronicle', p. 27.
23. E. Sivan, *L'Islam et la croisade* (Paris, 1968) pp. 70–2.
24. A. Morabia, 'Ibn Taymiyya, dernier grand théoricien du Gihad médiéval', *Bulletin d'études orientales*, **30** (1978) 85–99; C. Hillenbrand, *The Crusades: Islamic Perspectives* (Edinburgh, 1999) pp. 241–3. See also A. Morabia, *Le Gihad dans l'Islam médiéval* (Paris, 1983) *passim*.
25. P. Carey, 'The Origins of the Java War (1825–30)', *English Historical Review*, **91** (1976) 75–8.
26. H.-T. H. Tai, *Millenarianism and Peasant Politics in Vietnam* (Cambridge, MA, 1983) *passim*.
27. M. Adas, *Prophets of Rebellion* (Cambridge, 1979) *passim*.
28. For Innocent, see H. Roscher, *Papst Innocenz III und die Kreuzzüge* (Göttingen, 1969) pp. 288–91. For the First Crusade, see below.
29. E.-D. Hehl, *Kirche und Krieg im 12. Jahrhundert* (Stuttgart, 1980) p. 42.
30. Peter the Venerable, *The Letters*, ed. G. Constable, 2 vols (Cambridge, MA, 1967), vol. 1, p. 409.
31. 'Historia de expeditione Friderici imperatoris', ed. A. Chroust, MGH, Scriptores Rerum Germanicarum, nova series, no. 5 (Berlin, 1928) pp. 8–9.
32. Peter of Vaux-de-Cernay, *Hystoria Albigensis*, ed. P. Guébin and E. Lyon, 3 vols (Paris, 1926) vol. 1, pp. 64–5.
33. Innocent III 214.514.
34. See, for example, A.Theiner, *Vetera monumenta historica Hungariam sacram illustrantia*, 2 vols (Rome, 1859–60) vol. 1, pp. 178–9.
35. See S. D. Goitein, *A Mediterranean Society*, 6 vols (Berkeley, CA, 1967–93) vol. 2, pp. 273–407.
36. Morabia, 'Ibn Taymiyya', p. 94.
37. Peter the Venerable, *The Letters*, pp. 327–30.
38. For example, Bernard of Clairvaux, *Opera*, ed. J. Leclercq and H. M. Rochais, 8 vols (Rome, 1957–77) vol. 8, pp. 311–17; Peter the Venerable, *The Letters*, p. 328; Innocent III, in Simonsohn, *The Apostolic See: Documents*, pp. 86–7, 92–3.
39. Solomon bar Simson, 'Chronicle', p. 28.
40. Guibert of Nogent, *Dei Gesta per Francos*, ed. R. B. C. Huygens, CCCM 127A (Turnholt, 1996) pp. 113–14.
41. Riley-Smith, *The First Crusaders*, p. 62.

42. Ekkehard of Aura, 'Hierosolymita', RHC Oc 5, p. 20.
43. Richard of Poitiers, 'Chronicon', RHGF 12, pp. 411–12.
44. Guibert of Nogent, *De vita sua sive Monodiae*, 2.5, ed. and trans. E.-R. Labande, Les Classiques de l'Histoire de France au Moyen Age, vol. 34 (Paris, 1981) pp. 246–8.
45. See Riley-Smith, *The First Crusade and the Idea of Crusading*, p. 51, although the argument now requires some modification.
46. Guibert of Nogent, *De vita sua*, pp. 246–8.
47. Chazan, 'Emperor Frederick I', *passim*.
48. Matthew Paris, *Chronica maiora*, ed. H. R. Luard, 7 vols, RS 57 (London, 1872–83) vol. 5, pp. 361–2. See W. C. Jordan, *Louis IX and the Challenge of the Crusade* (Princeton, NJ, 1979) p. 154.
49. Louis IX in F. Duchesne, *Historia Francorum Scriptores*, 5 vols (Paris, 1636–49) vol. 5, p. 429.
50. P. Cole, D. L. d'Avray and J. S. C. Riley-Smith, 'Application of Theology to Current Affairs: Memorial Sermons on the Dead of Mansurah and on Innocent IV', *Historical Research*, **63** (1990) 227–39. The sermons have been edited by P. Cole in *The Preaching of the Crusades to the Holy Land, 1095–1270* (Cambridge, MA, 1991) pp. 235–43.
51. See N. P. Zacour, 'The Children's Crusade', in K. M. Setton (ed.-in-chief), *A History of the Crusades*, 2nd edn, 6 vols (Madison, WI, 1969–89) vol. 2, pp. 326–8.
52. See B. Z. Kedar, 'The Forcible Baptisms of 1096: History and Historiography', in K. Borchardt and E. Bünz (eds), *Forschungen zur Reichs- Papst- und Landesgeschichte: Peter Herde zum 65. Geburtstag von Freunden, Schülern und Kollegen dargebracht* (Stuttgart, 1998) vol. 1, pp. 187–200.
53. See J. A. Brundage, *Medieval Canon Law and the Crusader* (Madison, WI, 1969) pp. 30–114.
54. See, for example, Riley-Smith, *The First Crusade and the Idea of Crusading*, pp. 58–90.
55. Riley-Smith, *The First Crusaders*, pp. 54–60.
56. Eudes of Deuil, *De profectione Ludovici VII in orientem*, ed. V. G. Berry (New York, 1948) p. 8.
57. Baha' ad-Din, *Kitab al-nawadir al-sultaniya wa'l-mahasin al-yusufiya*, trans. C. W. Wilson (London, 1897) pp. 207–8.
58. A. Becker, *Papst Urban II*, 2 vols (Stuttgart, 1964–88) vol. 2, pp. 435–58.
59. 'Historia de expeditione Friderici', p. 14; Chazan, 'Emperor Frederick I', pp. 83–93.
60. S. Lloyd, *English Society and the Crusade, 1216–1307* (Oxford, 1988) pp. 55–6.
61. For example, the description of Oliver of Cologne (or Paderborn) of his preaching in 1214: *Schriften*, ed. H. Hoogeweg (Tübingen, 1894) p. 285.
62. 'Historia de expeditione Friderici', p. 6.
63. 'Quia maior', ed. G.Tangl, *Studien zum Register Innocenz' III* (Weimar, 1929) pp. 88–9.
64. On the homilies, see now C. T. Maier, *Crusade Propaganda and Ideology: Model Sermons for the Preaching of the Cross* (Cambridge, 2000) *passim*.
65. Humbert of Romans, chapters 1–26. For Humbert, see E. T. Brett, *Humbert of Romans: His Life and Views of Thirteenth-Century Society* (Toronto, 1984) pp. 3–11.

66. Gunther of Pairis, *Hystoria Constantinopolitana*, ed. P. Orth (Berlin, 1994) p. 113. See C. T. Maier, 'Kirche, Kreuz und Ritual: eine Kreuzzugspredigt in Basel im Jahr 1200', *Deutsches Archiv*, 55 (1999) 95–115.

67. Humbert of Romans, chapter 1.

68. For much later instructions about this, see N. Housley, *Documents on the Later Crusades, 1274–1580* (Basingstoke, 1996) p. 151.

69. Eudes of Deuil, p. 8.

70. In 1123 the bishops at the First Lateran Council decreed against those 'who had taken their crosses off' and had not left on crusade: *Conciliorum Oecumenicorum Decreta*, ed. J. Alberigo et al. (Freiburg, 1962) p. 168. For the crosses themselves, see Riley-Smith, *The First Crusaders*, pp. 11–12.

71. Baldric of Bourgueil, 'Historia Jerosolimitana', RHC Oc 4, p. 101.

72. James of Vitry, 'Sermones vulgares', ed. J. B. Pitra, *Analecta novissima*, 2 vols (Paris, 1885–8) vol. 2, pp. 419–20. James was reflecting the language of Innocent III, 'Quia maior', pp. 88–91.

2
The Anti-Jewish Violence of 1096: Perpetrators and Dynamics

Robert Chazan

The anti-Jewish passions unleashed during the early stages of the First Crusade raise a variety of intriguing issues. The persecution of Jews illuminates, in a painful way, the uncertainties of a new and diffuse movement, groping towards clarification of its constituency and objectives.[1] In the Jewish efforts at self-protection, we see a struggle, in the face of unanticipated upheaval, to reassert normalcy and to reimpose traditional ecclesiastical and secular safeguards. The unusual Jewish martyrdoms of 1096 show us the emergence of a fascinating Jewish counter-crusade mentality, with many of the beleaguered Jews of 1096 absorbing the vibrant emotionality evident in the crusading bands, adapting it towards traditional Jewish ideals, and cloaking it in the symbols of Jewish tradition.[2]

Addressing yet another interesting facet of the 1096 persecutions, the present paper will attempt to identify more closely the perpetrators of the anti-Jewish violence and to clarify the convoluted dynamic of anti-Jewish sentiment and behaviour.[3] This investigation is intended, in part, as a minor corrective to traditional views of the 1096 persecutions. More importantly, it offers a fascinating case study of the escalation of hostility, of the ways in which hatreds, once unleashed, spiral in wholly unanticipated directions.

It is generally assumed that the persecutions of 1096 were inflicted primarily, indeed solely, by popular crusading bands, for the most part German popular crusading bands. This view – while not rooted in the sources themselves, as we shall see – is widespread in modern crusade literature. To cite but one recent example of this assumption, let us note Sir Steven Runciman's portrayal of the anti-Jewish assaults of 1096. Runciman traces the movement of Count Emicho's army from Speyer through Worms through Mainz and on to the city of Cologne. He has a portion of Emicho's army detouring to Trier and on to Metz and then back to the Cologne area, where they assaulted the Jews gathered in Neuss, Wevelinghofen, Altenahr and Xanten. The only attacks not attributed by Runciman to Count Emicho were those at Prague, ascribed to Folcmar and his followers, and at Regensburg, assigned to Gottschalk and his troops. The picture is clear and consistent. The anti-Jewish excesses in 1096 can all be traced to one of three German crusading bands, with most of them perpetrated by Count Emicho and his horde.[4]

More precise identification of the perpetrators of the anti-Jewish violence is possible, given the combination of Jewish and Christian sources at our disposal. This clarification constitutes a minor issue only in the history of the First Crusade. More important is the light that such clarification can shed on the escalation of violence in 1096, on the ways in which a message of aggression and revenge was successively reinterpreted by ever-expanding circles into a multidimensional rationale for attack on Jews.

While the available Christian sources are of some help in specifying more precisely the perpetrators of anti-Jewish violence, it is the Hebrew narratives that are most closely attuned to the details of the 1096 assaults, including both the perpetrators of the attacks and their motivations. My recent work on the Hebrew First Crusade narratives has led me to identify five discernible voices available to us. The two earliest voices are the authors of the *Mainz Anonymous* and the Trier unit of the *Solomon bar Simson Chronicle*; considerably later are the author of the Cologne unit of the *Solomon bar Simson Chronicle*, the editor of the *Solomon bar Simson Chronicle*, and the editor of the *Eliezer bar Nathan Chronicle*. The earlier authors are far more oriented to the time-bound realities of 1096 and thus provide our most useful evidence on the perpetrators of the violence and their motivations.[5]

In order to make our way into the complexities of the anti-Jewish violence of 1096, let us begin by following the account of the early, sophisticated and meticulous Jewish author of the *Mainz Anonymous*.[6] His report will provide useful preliminary identification of the range of perpetrators involved in the 1096 attacks.

Our narrator pinpoints the origins of the First Crusade in northern France:

> It came to pass in the year 1028 after the destruction of the [Second] Temple that this disaster struck the Jews. For initially the barons and the nobles and the folk in France arose, conspired, and planned to go up, to rise like eagles, to wage war, to clear the way to Jerusalem the Holy City, and to reach the sepulchre of the crucified.[7]

Noteworthy here is the absence of any reference to papal instigation of the campaign. Our author was German and, in all likelihood, viewed the early stages of the enterprise from his own experience of northern-French crusaders crossing over into German territory; for such German Jews, the crusade was perceived largely as a northern-French initiative. To use the language of our perceptive German–Jewish observer, 'the barons and the nobles [note the conspicuous and accurate absence of royalty in the report] and the folk in France' set in motion the First Crusade.[8] The Jewish author shows sensitivity to the social complexity of the crusade, noting – as do most Christian sources – the involvement of both high-born and commonfolk. What united this motley combination of social groupings was a powerful religious vision, which the Jewish author, despite his denunciation of this vision, makes no effort to hide and, indeed, highlights.[9]

According to the *Mainz Anonymous*, the call to crusading almost immediately stimulated anti-Jewish sentiment as well:

> They [the crusaders] said to one another: 'Behold, we travel to a distant land to war against the rulers of that land. We take our lives in our hands in order to kill and to subjugate all the kingdoms that fail to acknowledge the crucified. How much more [should we be aroused against] the Jews, who killed and crucified him.'[10]

This distortion of the crusading call evoked considerable anti-Jewish hostility, plunging the Jews of northern France into deep anxiety. To be sure, the Jewish chronicler provides no information on anti-Jewish incidents in northern France.[11]

The author of the *Mainz Anonymous* goes out of his way to portray the rapid development of the First Crusade, highlighting the extent to which his fellow German Jews were utterly unaware of the dangers brewing further westward. The Jews of France, fully cognizant of the call to the crusade, were profoundly shaken and turned to their brethren in the Rhineland for their prayers and support. The Rhineland Jews are depicted as expressing grave fears for the well-being of their co-religionists in northern France, while ironically asserting their own lack of concern over any danger that might threaten them.

Fairly quickly, however, some of the French crusading armies began to make their way eastward through German territory. The French crusading forces that took this easterly route were very much in the minority, with most of the French militias opting for a more southerly path to the Holy Land. Nonetheless, the *Mainz Anonymous* presents an impression of the French forces that took the easterly route as enormous, 'battalion after battalion, like the army of Sennacherib'.[12] As these forces made their way across western Germany, the Jews whom they encountered contributed funds with which the crusading legions might provision themselves. The ploy was successful, and the *Mainz Anonymous* provides no evidence of violence perpetrated by any of the French militias as they made their way through German territory.[13]

According to the *Mainz Anonymous*, ever alert to the rapid evolution of the crusade and the emergence of new groupings and themes, unanticipated dangers were unleashed by the essentially peaceful passage of the French armies. The dangers developed from two quarters. On the one hand, the passage of French warriors stimulated crusading fervour among the German barons:

> As the [French] crusaders came, battalion after battalion, like the army of Sennacherib, some of the barons in this empire said: 'Why do we sit thus? Let us also go with them. For anyone who undertakes this journey and clears the way to ascend to the befouled sepulchre of the crucified [Jewish derogation of the Holy Sepulchre] will be assured hell [the Jewish author's inversion of Christian convictions of paradise].'[14]

Once again, our author indicates that crusading ardour gripped both the highborn and the lowly and that it immediately stimulated anti-Jewish sentiment. Just as the German crusaders were convinced that crusading would assure them paradise, so too were they persuaded of the rewards for anti-Jewish actions. 'They [the Germans attracted to the crusade] initiated a rumour throughout the Rhineland that anyone who would kill a Jew would have his sins forgiven. Indeed, there was one noble, named Ditmar, who said that he would not depart from this empire until he would kill a Jew. Then he would set forth.'[15] As had already happened in France, German crusaders – or at least some of them – drew damaging anti-Jewish inferences from the crusading call. According to the *Mainz Anonymous*, these inferences led immediately, in Germany, to random and spontaneous violence perpetrated by individual crusaders: 'The crusaders came with their insignia and their standards before our homes. When they saw one of us, they would run after him and pierce him with their spears, to the point that we were afraid to cross our thresholds.'[16] This is, to be sure, a portrait of sporadic and individualized violence, not the assault of an organized crusader army.

The *Mainz Anonymous* indicates that yet a second danger was unleashed by the passage of the French crusaders. 'Our sins brought it about that the burghers in every town through which the crusaders passed became contentious against us. Indeed, their [the burghers'] hand was with them [the crusaders] in destroying root and branch along all the way to Jerusalem.'[17] The *Mainz Anonymous*, deeply committed to precise identification of the perpetrators of the 1096 violence, highlights throughout the rest of its account the role played by burghers in the disaster that overtook Rhenish Jewry.[18]

Neither crusaders nor burghers were unanimous in perceiving anti-Jewish implications in the call to the crusade. Indeed, the majority of crusading armies show no evidence whatsoever of anti-Jewish sentiment or behaviour.[19] Likewise the burghers of western Christendom were by no means united in an anti-Jewish stand during the early months of the crusade. The *Mainz Anonymous* and the other Jewish sources testify as well to burgher friendliness and protectiveness towards endangered Jews.[20] What the *Mainz Anonymous* suggests, rather, is evocation of anti-Jewish sentiment in some of the crusading contingents and in segments of the burgher population all across German territory.

The only violence depicted to this point in the *Mainz Anonymous* was random and spontaneous, with no sense of significant Jewish casualties. There were two more serious kinds of aggression in the offing: that perpetrated by clusters of individual crusaders and burghers acting in concert, and that inflicted by an organized crusading army. Let us begin with the first. After his introductory observations on the generation of the First Crusade in northern France, the movement of French crusading bands into Germany, the arousal of German crusading fervour, and the evocation of burgher hostility, the author of the *Mainz Anonymous* focuses on three major Rhineland communities – the young Jewish settlement in Speyer and the two larger and older Jewries of Worms and Mainz.

The Speyer story shows us, for the first time in the *Mainz Anonymous*, a combination of crusaders and burghers working in tandem against Jews. The violence remained spontaneous. A coalition of crusaders and burghers hurriedly planned to surprise the Jews of Speyer at their Sabbath prayers; the Jews, alerted to the danger, prayed early and retreated to their homes; Bishop John intervened energetically to protect his Jews. Without in any way detracting from the impressiveness of the bishop's intentions and achievement, it is clear that the force he opposed (a spontaneous coalition of crusaders and burghers) was far less imposing than the forces subsequently massed against the Jews of Worms and Mainz. The chaos of the early stages of the crusade must be firmly borne in mind. While we tend to visualize the First Crusade as a series of organized military forces (the successes of 1099 were achieved by a set of such organized military forces), in its earliest stages individual crusaders and clusters of crusaders roamed about, seeking to find their place in organized contingents. It is precisely such individuals and clusters that linked themselves with Speyer burghers in planning the abortive assault on the synagogue of Speyer. What should be emphasized is the absence of any organized crusader army appearing from the outside and intruding into Speyer. Runciman's sense of the role of Emicho's army in the Speyer attack finds no support in our meticulous Jewish source. In fact, the *Mainz Anonymous* notes explicitly the subsequent passage of Emicho's army through the area in which the Speyer Jews were sequestered, thus pointedly highlighting its lack of involvement in the spontaneous crusader–burgher assault hurriedly mounted in Speyer itself.[21]

In Worms, the Jewish community split into two segments, with one group of Jews opting to remain in their homes and the second seeking refuge in the bishop's castle. A large-scale assault was mounted against the first group by yet another combination of crusaders and burghers, energized spontaneously by a successful ruse.

> It came to pass on the tenth of Iyyar, on Sunday, that they plotted craftily against them [the Jews of Worms who had remained in their homes]. They took a trampled corpse of theirs, that had been buried thirty days previously, and carried it through the city, saying: 'Behold what the Jews have done to our comrade. They took a Christian and boiled him in water. They then poured the water into our wells in order to kill us.' When the crusaders and burghers [note the combination] heard this, they cried out and gathered – all who bore and unsheathed [a sword], from great to small – saying: 'Behold the time has come to avenge him who was crucified, whom their ancestors slew. Now let not a remnant or a residue escape, not even an infant or a suckling in the cradle.'[22]

The mixed force of crusaders and burghers that coalesced in response to the ruse seems to have inflicted heavy casualties on the Worms Jews who had remained in their homes, with numbers of them accepting the option of conversion. This initial assault in Worms, still spontaneous, represents, from the Jewish perspective, the first major disaster associated with the crusade. Once again, there is no reflection in our account of the involvement of Count Emicho's army, or any other crusader army for that matter, in the initial assault in Worms.

More and worse was yet to come. The second group of Worms Jews was subjected to a more carefully orchestrated attack:

> It came to pass on the twenty-fifth of Iyyar that the crusaders and the burghers said: 'Behold those who remain in the courtyard of the bishop and his chambers. Let us take vengeance on them as well.' They gathered from all the villages in the vicinity [an unusual reference to villager participation in the anti-Jewish assaults], along with the crusaders and the burghers; they besieged them [the Jews]; and they did battle against them. There took place a very great battle, one side against the other, until they [the Christians] seized the chambers in which the children of the sacred covenant were.[23]

With the Jews defeated in their effort to hold off the attackers, the fate of this segment of Worms Jewry was sealed. In more premeditated and orderly fashion than any case depicted heretofore, the second assault on Worms Jewry shows us a force of crusaders and burghers with considerable military power. Jewish defences crumbled in the face of this assault, and the second group of Worms Jews lay exposed to near-total destruction.

Thus, to this point, we have seen in the *Mainz Anonymous*: anti-Jewish hostility evoked in crusader ranks; related anti-Jewish animus in segments of the burgher population of the Rhineland towns; random and individualized crusader violence; crusader–burgher aggression spawned spontaneously, with at least one instance – the opening assault in Worms – in which this aggression resulted in significant Jewish casualties; an organized and premeditated crusader–burgher coalition, again in Worms, committed to total destruction of a large group of well-defended Jews. At no point in the report on events in Speyer or Worms, however, does a formal crusading army make an appearance. In the *Mainz Anonymous's* careful account, it is only in Mainz that a further and most potent source of anti-Jewish violence emerges – a crusading army ideologically committed to the annihilation of Jews, either through conversion or killing.

The *Mainz Anonymous's* account of the destruction of Mainz Jewry differs from all its prior reporting. In this account, a crusading army, that of Count Emicho, for the first time initiates and carries out the anti-Jewish violence. On the twenty-fifth of May, Count Emicho encamped his army outside the walls of Mainz, prepared to lead his troops in a military assault on the large and distinguished Jewish community of that town. For two days this crusading army remained outside the city walls, that had been bolted against it. On the third day, the gates to the city were opened by sympathetic burghers, and Emicho's army marched into Mainz.

Every aspect of the *Mainz Anonymous* depiction of the annihilation of Mainz Jewry differs from its earlier reportage. There was nothing spontaneous or ad hoc about the Mainz assault. It was carried out by an organized military force that camped outside the town, that prepared itself to storm the closed gates of the town walls, that was contacted by the endangered Jews but refused negotiation, that gained entry into the town through the collaboration of sympathetic burghers. Once inside Mainz, the army of Count Emicho made its way directly to the central refuge of Mainz Jewry. 'They [the army of Emicho] came with their standards to the gate of the archbishop, where the children of the sacred covenant were; [they constituted] a great army, like the sands on the seashore.'[24] This crusading army joined battle with the Jewish defenders of the archbishop's compound (the archbishop's militia had fled); victorious, its members invaded the vanquished fortification and slaughtered the Jews found throughout the compound.

Upon completion of this slaughter, the army of Emicho then made its way to a secondary enclave of Jews, in the burgrave's compound:

> Then the crusaders began to exult in the name of the crucified. They lifted their standards and came to the rest of the community, to the courtyard of the nobleman, the burgrave. They likewise besieged them [the Jews assembled in the burgrave's compound] and waged war against them. They seized the entranceway to the gate of the courtyard and smote them [the Jews].[25]

This description again projects a crusading force operating in military fashion, in sharp contrast to what had been described previously by the *Mainz Anonymous*. It seems likely that this last kind of violence (that perpetrated by an organized military force moved by radical commitment to destruction of Jews) was the costliest suffered by those Jews caught up on the margins of the early crusading movement.

Focusing on the best extant source for the anti-Jewish violence of 1096 has provided us with a number of useful insights. The *Mainz Anonymous* has, first of all, disabused us of the notion that all the anti-Jewish assaults were perpetrated by invading crusader forces. The portrait we have seen in Runciman, for example, is simplistic and inaccurate.[26] The reality was far more complex. In a more positive vein, the *Mainz Anonymous* helps to establish a framework for surveying other sources, both Jewish and Christian, in an effort to clarify further the identity of the perpetrators of the 1096 violence and the dynamic of escalating animosities.

The first useful step in pursuing our investigation is to isolate those incidents for which data have been preserved.[27] The locales of anti-Jewish violence which have left traces and the sources available to us include: Rouen – Guibert of Nogent;[28] Speyer – the *Mainz Anonymous*, the *Solomon bar Simson Chronicle* and Bernold of St Blaise;[29] Worms – the same two Hebrew narratives and Bernold;[30] Mainz – the same two Hebrew narratives, Albert of Aachen, the *Annalista Saxo* and the *Annales Wirziburgenses*;[31] Cologne – the *Solomon bar Simson Chronicle* and Albert of Aachen;[32] Trier – the *Solomon bar Simson Chronicle* and the *Gesta Treverorum*;[33] Metz – the *Solomon bar Simson Chronicle*;[34] Regensburg – the *Solomon bar Simson Chronicle*;[35] Prague – Cosmos of Prague and the *Annalista Saxo*.[36]

With respect to the possibility that information on major attacks might have been lost, two points are notable. The first is that the so-called *Solomon bar Simson Chronicle* is in fact a compilation of materials from the mid-twelfth century, a compilation that includes even a hazy, unlikely story about Jews being saved in some unidentifiable place.[37] That major further persecutions took place, of which the compiler was unaware, seems unlikely. Moreover, the most devastating assaults – Worms, Mainz and Cologne – are all documented in more than one source, reinforcing the sense that we are unlikely to be deprived of information on major 1096 episodes.

Our examination of the *Mainz Anonymous* suggested that the very worst violence was that committed in Mainz, in premeditated and military fashion, by the organized crusading band of Count Emicho.[38] It seems likely that the same band was responsible for the effort to wipe out the enclaves of Cologne Jews who had been placed by their bishop in rural redoubts for safety. The archbishop of Cologne had emulated the successful example of the bishop of Speyer. He distributed his Jews among a number of outlying fortifications, clearly in hopes that he would thereby diminish the incentive to military assault. The ploy was an intelligent one; it misassessed, however, the radical determination of Count Emicho and his followers to destroy Cologne Jewry.

Although the author of the Cologne unit of the *Solomon bar Simson Chronicle* is uninterested in specifying the perpetrators of the anti-Jewish violence that overtook almost all the enclaves of Cologne Jews, his descriptions leave no doubt as to the reality of an organized crusading force. In refuge after refuge, outside troops are described as descending upon the endangered Jews. Thus, the assault on the Jews sequestered in Wevelinghofen is depicted in the following terms: 'When the enemies came before the town, then some of the saintly went up on the tower and threw themselves into the Rhine River that circles around the town.'[39] Note the enemy coming towards the town. At Altenahr: 'On the fourth of the month of Tammuz, a Thursday, the enemies gathered against the saintly of Altenahr, to torture them with great and terrible tortures until they [the Jews] might agree to baptism.'[40] The description of the arrival of the crusaders in Xanten on the eve of the Jewish Sabbath is particularly revealing: 'On Friday, the fifth of the month, on the eve of the Sabbath, at dusk on the eve of rest, the enemy – the enemy of the Lord – came upon the saintly of Xanten. The enemies arose against them at the hour of the sanctification of the [Sabbath] day. They had seated themselves to break bread. They had sanctified the [Sabbath] day with the prayer *Va-yekhulu* and had made the benediction over the bread. They then heard the sound of the oppressor; the seething waters came upon them.'[41] Perhaps the clearest testimony of all is provided by the fate of the Cologne Jews gathered in Moers. There, a siege is indicated, with the local authorities negotiating with the crusading force in an effort to avoid bloodshed within Moers and against its Jews. All through the account of successful negotiation and eventual baptism of almost all the Jews found in Moers, the presence of an outside crusading force is highlighted.[42]

Given the dating of these assaults during the closing days of the month of June, their proximity in time to Count Emicho's assault on Mainz, and the parallel commitment to complete obliteration of Jewish life either through baptism or slaughter, we are warranted in concluding that the destruction of both Mainz and Cologne Jewry can be attributed to the radical thinking and behaviour of the crusading band of Count Emicho. These are, however, the only assaults attributable to the crusading force of Count Emicho.[43]

Another instance of violence committed directly by a band of crusaders is the poorly detailed attack on the Jews of Prague. It seems highly unlikely that the casualties of this third documented crusader assault were heavy. Our sources, Cosmos of Prague and the *Annalista Saxo*, highlight in their sketchy reports conversion and the subsequent reversion of the converts to Judaism. Rather clearly, however, the attack was carried out by a crusader contingent making its way eastward in organized fashion.[44]

Thus, the number of attacks perpetrated directly by crusader bands on the march was quite limited, although the casualties in Mainz and

Cologne were considerable, accounting for a significant percentage of the total of Jewish lives lost in 1096.[45] The limited number of attacks leaves us with one remaining question, and that involves the ebb and flow of such assaults. If Count Emicho and his followers were so intensely committed to total destruction of the Jews of Mainz and the dispersed Jews of Cologne, why then did they not engage in further assaults as they made their way eastward? This is a question for which I have no sure answer at the moment. Again, it seems highly unlikely that Emicho's forces committed major violence that has gone unrecorded. It may well be that the anti-Jewish sentiment within Emicho's army was aroused largely by the great Jewish communities of the Rhineland and that the lesser Jewish settlements further eastward simply went more or less unnoticed. In any case, the reality of a limited number of attacks by organized crusader armies seems incontrovertible.

Turning back once more to the *Mainz Anonymous*, we recall that the next most costly type of attack took place in Worms, involving a coalition of crusaders, burghers and villagers operating in premeditated fashion. Again, there is no sense that the crusaders noted here constituted an identifiable band moving on its way eastward; the sense is, rather, of a random group of crusaders operating in an ad hoc way with their burgher and villager allies. Nonetheless, the premeditated combination of unorganized crusaders, burghers and villagers was a potent one, and Jewish casualties in this second assault on Worms Jewry ran high. We have also noted spontaneous coalitions of crusaders and burghers, responsible for the killing of those Worms Jews who had opted to remain in their homes and for the abortive assault on the Jews of Speyer.

Another attack that seems to show the same profile – a combination of unorganized crusaders and burghers operating together in either premeditated or spontaneous fashion – seems to have taken place in Regensburg. The report on the violence in Regensburg included in the *Solomon bar Simson Chronicle* is sketchy in the extreme. It simply indicates the following: 'The [Jewish] community that was in Regensburg was converted in its entirety, for they saw that they could not be saved. Moreover, when the crusaders and common-folk gathered against them [the Jews], those who were in the town [seemingly a reference to the burghers of Regensburg] forced them against their will into a certain river and made the evil sign over the water – the cross – and baptized them all in unison in that river.'[46] This is a most difficult passage to interpret. It could conceivably refer to a premeditated crusader–burgher coalition or to a spontaneous crusader–burgher coalition. In any case, there were, according to our report, no Jewish deaths in Regensburg,

with the entire Jewry converted against its will and subsequently returning to the Jewish fold.

There is yet one more instance of crusader–burgher cooperation in
our sources, and that involves the Jewish community of Trier. The Trier
story is available to us in a discrete segment of the composite *Solomon
bar Simson Chronicle*. I have dated this narrative fairly early and have highlighted its focus on explaining the conversion of Trier Jewry by recon-
structing carefully the evolution of protracted pressure on the Jews of
that town.[47] The result, for our purposes, is a fairly detailed description
of diverse stages in the persecution of Trier's Jews, although this
detailed description lacks the elegant clarity of the *Mainz Anonymous*.
For the moment, let us focus only on the denouement – the period after
Pentecost, during which a combination of crusaders and burghers
besieged the Jews, who had found refuge in the bishop's palace. Once
again, we encounter the same combination of crusaders and burghers,
this time involved with the local bishop in complex negotiations that
would eventuate in the mass conversion of Trier Jews.

Thus, the assaults of 1096 were not perpetrated exclusively by popular crusading armies. The situation is considerably more complex. On
occasion, crusading armies were involved in the attacks, usually with
deadly results. At other times, coalitions of individual crusaders or clusters
of crusaders and hostile burghers formed, with the objective of attacking
local Jews. To the extent that these coalitions were organized, they too
could have deadly results; to the extent that they were spontaneous, the
results tended to be more limited. In a general way, greater attention
must be paid to the evocation of burgher antipathy to Jewish neighbours.

Indeed, there are a number of cases in which burghers were solely
responsible for assaults on Jews in 1096. In the town of Cologne, there
is interesting evidence of burgher violence, triggered by reports of the
massacre in Mainz. The reports of the Hebrew and Latin narratives are
fully consistent one with the other. According to the Cologne segment
of the *Solomon bar Simson Chronicle*,

> when they [the Jews of Cologne] heard that the communities had
> been killed [seemingly a reference to events in Worms and Mainz],
> the Jews fled to their Christian acquaintances. They remained there
> during the two days of the holiday [the two days of Shavuot, 30–1
> May]. On the third day, in the morning, there was thunderous noise,
> and the enemy arose against them, broke into their homes, plun-
> dered, and took booty. They destroyed the synagogue, took out the
> Torah scrolls, and desecrated them, trampling them in the streets.[48]

The Hebrew account then notes three instances of Jewish loss of life, in all cases involving Jews who had either spurned or left the protection of their Christian neighbours.

As obvious in the above citation, the author of the Cologne segment of the *Solomon bar Simson Chronicle* was uninterested in identifying the perpetrators of the anti-Jewish violence. Here, as was generally the case, he was satisfied to merely label them the enemy. The emphasis on booty, however, suggests a somewhat different group of attackers from that operating in Mainz. The Latin account of Albert of Aachen, which identifies Count Emicho as responsible for the massacre in Mainz, clarifies the situation in Cologne, specifying the perpetrators of the initial assaults as burghers. 'This slaughter of Jews was done first by burghers of Cologne. These suddenly fell upon a small band of Jews and severely wounded and killed many; they destroyed the houses and synagogues of the Jews and divided among themselves a very large amount of money.'[49] There is striking convergence between the two descriptions, leaving little room for doubting that the burghers of Cologne were responsible for the initial violence in that city.[50]

While the final violence in Trier involved a coalition of crusaders and burghers, the Trier segment of the *Solomon bar Simson Chronicle* indicates pure burgher anti-Jewish violence as well. Early in his report, the anonymous author of this Trier narrative notes the arrival of Peter the Hermit – whom he identifies specifically – and his minions. He indicates that Peter bore with him a letter from the Jews of France, advising financial support of Peter and his followers, in return for which any potential violence would be suppressed. The Jews of Trier followed the advice of their French brethren, and Peter's army left without incident. Thus, the brief report of the *Mainz Anonymous* on the peaceful passage of the French crusading bands is well fleshed out by the fuller description of the movement of Peter the Hermit and his followers through Trier. Indeed, both sources further agree that Jewish hopes that the passage of the French armies would spell the end of danger were quickly dashed.

At this point, when the Jews of Trier might have considered themselves fortunate to have successfully escaped danger, it quickly became clear that such was not in fact the case. Even while describing the passage of Peter through Trier, the author of the Trier unit introduces the burghers of that town, at first in what seems to be a curious aside: 'When he [Peter] came here, our souls departed and our hearts were broken; trembling seized us and our holiday [Peter arrived on the first day of Passover] was turned into mourning. For to this point the burghers did not intend to inflict any harm on the [Jewish] community, until those pseudo-saintly ones arrived.' Now, the depiction of Jewish fears is certainly understandable, but one would assume that these anxieties involved Peter and his forces. Reference to the burghers at this juncture is strange, reflecting understanding conferred by hindsight. Immediately after noting the provisioning of Peter and his peaceful departure eastward, the author of the Trier unit continues: 'Then our wicked neighbours, the burghers, came and were envious of all the happenings that had befallen the rest of the [Jewish] communities in the land of Lotharingia. They [the burghers of Trier] heard what had been done to them and what had been decreed for them – great tragedy.'[51]

The Trier unit describes an invasion of the synagogue of Trier, with desecration of its *sancta* in a manner reminiscent of what occurred in Cologne. Once again, the perpetrators of this invasion and desecration seem to have been Trier burghers, since this incident seems to have taken place some time prior to the emergence of the crusader–burgher coalition that brought about the mass conversion.[52]

The Cologne segment of the *Solomon bar Simson Chronicle* provides, in addition, evidence of a villager eruption of violence, against a Jewish family that happened to fall into the clutches of this group of rural folk. The sad tale of a Cologne Jew named Shmaryahu, his wife and their three children is appended to the story of the forced conversion of the Cologne Jews who had sought refuge in Moers. This Jewish family, according to the Hebrew report, was taken in hand by an official of the bishop, who led them about, while they attempted to gain funds from their grown sons, who were safe in Speyer. When the money finally reached Shmaryahu and his protector, the latter betrayed them and delivered them into the hands of a group of villagers in a place that can no longer be identified.

These villagers, who knew the Cologne Jew, were delighted with the possibility of converting him and his family. The Jews feigned willingness for conversion, but, during the night, Shmaryahu took the lives of his wife and children, failing, however, in the effort to kill himself. Found alive in the morning alongside his dead spouse and offspring, Shmaryahu proclaimed himself ready for martyrdom, but the villagers withheld a simple and straightforward death. Instead, they buried the Jew alive, alongside the already dead members of his family, extricating him periodically in hopes of winning him to conversion. Eventually, the torture took its toll, and he expired.[53]

Thus, the perpetrators of the 1096 violence were diversified – organized crusading bands; random crusaders; random crusaders in concert with burghers; burghers in concert with random crusaders; burghers acting on their own; even occasionally villagers. The time has come to attempt some observations on the motivations of these diverse groupings. In its simplest form, the question to be posed is: How did the papal call at Clermont, which set the First Crusade in motion, eventuate in the anti-Jewish actions of the burghers of Cologne and Trier or the perverse cruelties impose on the Jew Shmaryahu by a group of Rhineland villagers?

Let us begin with Clermont itself. It is agreed that our extant sources do not permit reliable reconstruction of the papal appeal that launched the First Crusade.[54] Nonetheless, it seems clear enough that the momentous papal call included no direct reference to crusader behaviour towards the Jews. More precisely, it is almost certain that Pope Urban II called for no anti-Jewish violence, condoned no aggression upon the Jews, suggested no financial exploitation of the Jews in support of the crusade, or – by contrast – demanded no special protection for European Jewry.

At least two considerations militate against the possibility of either papal call for anti-Jewish violence or even papal sanction of aggression upon the Jews. The first is that such a call or even such a sanction would contravene well-established ecclesiastical guidelines for treatment of Jews; moreover, the behaviours of most of the main crusading armies, which included no anti-Jewish activities, hardly reflect either such a papal call or such papal acquiescence. Special mention of the Jews in a protective tone is similarly unlikely. All evidence suggests that Christians and Jews alike were taken utterly by surprise at the escalating anti-Jewish sentiments and behaviours. That the pope had been aware of such possibilities and others so totally unaware strains credulity. Moreover, when Bernard of Clairvaux, on the eve of the Second Crusade, marshalled a series of potent arguments against anti-Jewish violence by crusaders, he conspicuously omitted any papal policy statement promulgated in the early stages of the First Crusade.[55] We may be quite certain that the papal speech, whatever it might have included, made no reference of any kind to the Jews.

To be sure, there were motifs in the call to the crusade that lent themselves to distortion, that provided some basis for the appeal to anti-Jewish actions. Twentieth-century historians have laboured hard to identify the key elements in early crusade preaching and have succeeded admirably.[56] Three core crusading motifs lent themselves to anti-Jewish distortions. The first of the three is the notion of holy war, leading to the suggestion that, of all enemies, the Jews are the most despicable. There is a related crusading rationale that was probably more readily alterable into the basis for anti-Jewish action, and that was an emphasis on revenge against an enemy that had inflicted harm on Christianity and Christians. This might well be seen as simply a subset of the broader notion of war against the non-believing foe, but it deserves special mention.[57] Finally, most recent historians agree on the centrality of Jerusalem to First Crusade thinking. More specifically, it was the Holy Sepulchre, the most sacred of the Christian symbols associated with Jerusalem, that fired the crusader imagination and lent such emotional dynamism to the difficult enterprise. Now, the imagery of the Holy Sepulchre had especially negative implications for the Jews. For those setting forth to avenge alleged Muslim misdeeds relative to the Holy Sepulchre, the prior and more heinous purported misdeeds of the Jews, who had, in effect, created this central Christian shrine, had to demand more immediate redress.[58]

This set of three related crusade motifs each held the potential for generating anti-Jewish sentiment. The combination of the three probably enhanced the likelihood of anti-Jewish thinking. Once more, this is not to suggest that the anti-Jewish inferences were widely drawn among the crusading bands that set forth from western Christendom – they were not. Rather, crusade thinking merely bore a certain level of potential, activated in fairly restricted crusader circles.

How then are we to account for the related burgher violence of 1096, carried out sometimes in conjunction with crusaders and sometimes in purely burgher assaults? Unfortunately, our sources are far less revealing with respect to burgher antipathy than they are with respect to the crusaders.[59] Nonetheless, we may tentatively consider a number of likely possibilities. In the first place, the stirring of Christian hearts and minds to anti-Muslim – and by extension anti-Jewish – animosity was probably not confined to crusaders only. As charismatic preachers made their appeals to the warrior class, the message of hatred and violence had to be absorbed by others. Burghers not participating in the march eastward may have shared the general exhilaration spawned by the enterprise, and some of its anti-Jewish distortions as well.

The impact of general crusade thinking on the urban populace seems to be reflected in the fullest probing we have of the arousal of burgher animosity. As noted, Peter the Hermit and his forces made their way through Trier, were supported with provisions by the Jews of that town, and passed on without incident. The author of our anonymous account is quick to note, however, the immediate arousal of anti-Jewish sentiment in a burgher population heretofore friendly and peaceful. Once again, our anonymous Jewish chronicler is less specific than we might wish. He does seem to suggest, however, that the burghers of Trier were, in part, moved by some of the crusader thinking.[60]

At the same time, our Trier Jewish chronicler points in another direction as well. The passage of Peter and the Jewish financial contribution to his enterprise illuminated the vulnerability of the Jews and their fiscal resources. Immediately after noting the providing of funds to Peter the Hermit and the peaceful departure of his troops, our author continues: 'They [the Jews of Trier] took their money and bribed the burghers, each one individually. All this was unavailing on the day of the Lord's anger.'[61] Thus, exploitation of this vulnerability and these resources became an appealing option for the burgher neighbours of the Trier Jews. 'Spiritual' considerations were supplemented by 'material' considerations as well.

In a more general way, we know that tensions between Christian burghers and their Jewish neighbours predated the call to the crusade. Speyer Jewry, recurrently noted, was founded in 1084, in part out of the desire of some Mainz Jews to escape the threat of persecution at the hands of the latter town's burghers. As part of his effort to settle these endangered Jews in Speyer, Bishop Rudiger proposed to build a wall around the new Jewish enclave, in order to protect his Jews from the animosities of their burgher neighbours.[62] To be sure, pre-1096 burgher animosity towards Jews was hardly a constant. Indeed, the author of the anonymous account of the fate of Trier Jewry, victims of specifically burgher animosity, goes out of his way to indicate that, prior to the arrival of Peter the Hermit and his followers, relations between the Jews and their burgher neighbours had been perfectly peaceful. Nonetheless, in some instances at least, the intrusion of crusaders and crusade emotion intensified pre-existent burgher anti-Jewish sentiment and provided a convenient pretext for the squaring of old scores.

Finally, there is the looser and more general contagion of hostility and violence, the tendency all too frequently documented for hostility to beget further hostility and for violence to beget further violence. This general tendency seems to flow in large measure from the removal of normal societal constraints. As hatred and violence sweep across a society, the dismantling of normal inhibitions and the arousal of wide-ranging animosities often seem to go hand in hand.

In the case of the 1096 assaults, we can in fact identify something akin to an ideology of proliferating violence, the sense that evidence of anti-Jewish violence in itself serves as justification for further violence. The core of this rationalization of anti-Jewish behaviour involved divine favour. Recurrently, Jews were confronted with the argument that catastrophe itself served as evidence of God's abandonment of the Jewish people. For Christians advancing this argument, the only feasible

Jewish option in the face of allegedly obvious divine abandonment was conversion, with death the alternative to baptism.

Let us note but one example of this line of thinking. After depicting the second assault on Worms Jewry in general terms, the *Mainz Anonymous* proceeds to detail a number of specific instances of Jewish martyrdom. The last in this sequence of illuminating and moving stories concerns a Jewess named Minna, who had been hidden outside the town by Christian friends. After protecting her through the periods of actual assault, her Christian friends subsequently turned upon her, seemingly moved by the notion that the carnage in Worms served as irrefutable proof of divine rejection, in the face of which surviving Jews had no reasonable option other than baptism. These strangely sympathetic protectors implored her to convert. In the face of her adamant rejection of baptism, these erstwhile friends were moved to put this Jewess to death.[63] While the logic of this argumentation does not make a great deal of rational sense, some Christians saw in the violence itself justification for further violence. It is this sort of reasoning that seems to have moved the villagers cited earlier to force baptism and then death upon the Cologne Jew Shmaryahu and his family.

Ultimately, the assaults of 1096 provide an unusual case study of the dynamic of hatred and persecution. The call to the crusade, seemingly addressed to the professional military class of western Christendom and targeting a specific and far-off enemy, resonated in unanticipated ways within Europe itself. Ever broader circles of Christians attached themselves to the crusading enterprise, and the call to battle and revenge led some crusaders into anti-Jewish behaviour, as the imagery spawned by the crusade was interpreted in increasingly loose fashion. Eventually, the anti-Jewish motifs seem to have moved burghers and villagers who were not part of the crusading effort. Inevitably, material considerations and prior animosities played into the anti-Jewish equation as well. Ultimately, sanctioned violence seems to have a tendency to beget further and unsanctioned violence, a phenomenon observable in all ages.

In this early stage of the undertaking, the Church had not yet established firm guidelines for conducting the crusade; in any case, ecclesiastical leadership had only the barest control of the forces it had set in motion. A campaign rooted in religious zeal and hatred attracted unexpected numbers to its call, influenced profoundly even those who did not become part of it, ignited animosities against an expanding circle of perceived enemies, aroused passions to exploit perceived weakness and settle old scores, and ultimately took unanticipated victims.

Notes

1. There is a vast literature on the thinking, ecclesiastical and lay, that under-girded the First Crusade. Much of this thinking has been heavily influenced by the outstanding 1935 study of Carl Erdmann, now available in English as *The Origin of the Idea of Crusade*, trans. M. W. Baldwin and W. Goffart (Princeton, NJ, 1977). The most recent overview of these matters is J. Riley-Smith, *The First Crusade and the Idea of Crusading* (London, 1986). Particularly valuable for the lay thinking is the innovative study of M. Bull, *Knightly Piety and the Lay Response to the First Crusade: The Limousin and Gascony c. 970–c. 1130* (Oxford, 1993).

2. I have studied the assaults of 1096 and the variety of Jewish reactions they elicited in *European Jewry and the First Crusade* (Berkeley and Los Angeles, CA, 1987) and reformulated my findings in a more popularly oriented account entitled *In the Year 1096 . . . : The First Crusade and the Jews* (Philadelphia, PA, 1996).

3. While I indicated in my afore-cited treatments of the 1096 violence the involvement of both crusaders and burghers, I made no effort to study in depth the dynamics of anti-Jewish behaviour and thinking.

4. S. Runciman, *A History of the Crusades*, 3 vols (Cambridge, 1951–67) vol. 1, pp. 134–41. Such recent surveys of crusade history as K. M. Setton (ed.), *A History of the Crusades*, 6 vols (Madison, WI, 1969–89); H. E. Mayer, *The Crusades*, trans. J. Gillingham (Oxford, 1972); J. Riley-Smith, *The Crusades: A Short History* (London, 1987), all portray the persecution of Jews in the same way, as perpetrated by popular German crusading bands. Indeed, all these accounts can be readily traced back to the influential catalogue of events composed by H. Hagenmeyer, 'Chronologie de la premiere croisade', *Revue de l'Orient latin*, 6 (1898) 214–93 and 490–549.

5. I have recently completed a book-length study of the Hebrew First Crusade narratives entitled *God, Humanity, and History: The Hebrew First Crusade Narratives* (Berkeley and Los Angeles, CA, 2000). The analysis distinguishes between time-bound objectives (such as warning of crusading danger and identification of proper Jewish responses) and timeless objectives (such as rationalizing the violence and losses). Not surprisingly, it is the earlier accounts – the *Mainz Anonymous* and the Trier segment of the *Solomon bar Simson Chronicle* – in which the time-bound objectives are manifest. For that reason, I shall utilize most heavily these two voices in the present context.

6. For a full discussion of the *Mainz Anonymous*, see Chazan, *God, Humanity, and History*, Chapter 2.

7. The three Hebrew narratives were published in a scholarly edition by Adolf Neubauer and Moritz Stern (henceforth Neubauer and Stern), *Hebräische Berichte über die Judenverfolgungen während der Kreuzzüge* (Berlin, 1892), and were re-edited by Abraham Habermann (henceforth Habermann), *Sefer Gezerot Ashkenaz ve-Zarfat* (Jerusalem, 1945). An English translation of all three narratives can be found in Shlomo Eidelberg (henceforth Eidelberg), *The Jews and the Crusaders* (Madison, WI, 1977). Translations of the *Mainz Anonymous* and the *Solomon bar Simson Chronicle* can be found as an appendix to my *European Jewry and the First Crusade* (henceforth Chazan). This citation can be found in Neubauer and Stern, p. 47; Habermann, p. 93; Eidelberg, p. 99; Chazan, p. 225.

8. In the *Solomon bar Simson Chronicle*, there is brief and unfocused reference to the pope in this narrative's reworking of the earlier *Mainz Anonymous* account of the Speyer–Worms–Mainz episodes – see Neubauer and Stern, p. 4; Habermann, p. 27; Eidelberg, p. 26; Chazan, p. 248. This brief reference is clearly late and maladroit.

9. For more on this vision – and its distortions – see below.

10. Neubauer and Stern, p. 47; Habermann, p. 93; Eidelberg, p. 99; Chazan, p. 225.

11. In fact, Guibert of Nogent's autobiography – Guibert of Nogent, *De vita sua sive Monodiae*, ed. and trans. E.-R. Labande, Les Classiques de l'Histoire de France au Moyen Age, vol. 34 (Paris, 1981) pp. 246–8 (Latin) and pp. 247–9 (French trans.) – provides the only evidence we possess of crusade-related violence against the Jews of France.

12. Neubauer and Stern, p. 48; Habermann, p. 94; Eidelberg, p. 100; Chazan, p. 226.

13. The *Mainz Anonymous* suggestion of Jewish contribution of provisions and peaceful passage of the French crusaders is fleshed out by the Trier unit's fuller portrait of the peaceful passage of Peter the Hermit and his followers through Trier, which will be discussed below.

14. Neubauer and Stern, p. 48; Habermann, p. 94; Eidelberg, p. 100; Chazan, p. 226. For an important discussion of the use of such derogatory language in the Hebrew 1096 narratives, see A. Sapir Abulafia, 'Invectives against Christianity in the Hebrew chronicles of the First Crusade' in P. W. Edbury (ed.), *Crusade and Settlement* (Cardiff, 1985) pp. 66–72.

15. Neubauer and Stern, p. 48; Habermann, p. 94; Eidelberg, p. 100; Chazan, p. 226.

16. Neubauer and Stern, p. 48; Habermann, p. 94, Eidelberg, p. 100; Chazan, p. 227.

17. Neubauer and Stern, p. 48; Habermann, p. 94; Eidelberg, p. 100; Chazan, p. 226.

18. See, e.g., Neubauer and Stern, p. 48 (Speyer), p. 49 (the first assault in Worms), p. 49 (the second assault in Worms), p. 51 (the pre-Emicho threat of violence in Mainz), p. 53 (Emicho's entry into Mainz), p. 53 (the assault on the archbishop's palace in Mainz), p. 56 (the assault on David the *gabbai* and his family in Mainz); Habermann, pp. 94, 95, 96, 98, 99, 100, 103; Eidelberg, pp. 100, 102, 103, 106, 108, 109, 113; Chazan, pp. 227, 228, 229, 233, 235, 236, 241.

19. There are three extant Latin narratives composed by eyewitnesses to the First Crusade: (1) the anonymous *Gesta Francorum*, ed. and trans. R. Hill (Oxford, 1962); (2) Fulcher of Chartres, *Historia Hierosolymitana*, ed. H. Hagenmeyer (Heidelberg, 1913), trans. F. R. Ryan (Knoxville, TN, 1969); (3) Raymond of Aguilers, *Historia Francorum*, ed. J. H. and L. L. Hill (Paris, 1969), trans. J. H. and L. L. Hill (Philadelphia, PA, 1968). None of the three includes references to anti-Jewish violence committed by the armies whose exploits they recorded.

20. Note, for example, the well-known incident of the passage of a crusader group through Mainz, the taunting of Jews and escalation towards violence, and the protective stance adopted by a number of Mainz burghers. See Neubauer and Stern, pp. 51–2; Habermann, p. 98; Eidelberg, p. 106; Chazan, p. 233.

21. Neubauer and Stern, p. 48; Habermann, p. 94; Eidelberg, p. 101; Chazan, p. 227.
22. Neubauer and Stern, p. 49; Habermann, p. 95; Eidelberg, p. 102; Chazan, p. 228.
23. Neubauer and Stern, pp. 49–50; Habermann, p. 96; Eidelberg, p. 103; Chazan, pp. 229–30.
24. Neubauer and Stern, p. 53; Habermann, p. 99; Eidelberg, p. 108; Chazan, p. 235.
25. Neubauer and Stern, p. 55; Habermann, p. 102; Eidelberg, p. 112; Chazan, pp. 239–40.
26. Again, recall the dependence of Runciman and others on Hagenmeyer – see above, n. 4.
27. I shall only cite those sources that provide specific information on anti-Jewish incidents in given locales. In a general way, we shall note recurrently that the Jewish sources tend to be far fuller than the Christian. Of the three Hebrew narratives, I shall cite only the *Mainz Anonymous* and the *Solomon bar Simson Chronicle*, out of my sense that the narrative penned by Eliezer bar Nathan is wholly dependent on the latter. I have also argued in *God, Humanity, and History* that the Speyer–Worms–Mainz section of the *Solomon bar Simson Chronicle* is based on the *Mainz Anonymous*. However, the later narrative adds significantly to its source and will thus be cited separately.
28. Guibert de Nogent, *De vita sua*, pp. 246–8 (Latin) and pp. 247–9 (French trans.).
29. Neubauer and Stern, pp. 48 and 2; Habermann, pp. 94–5 and 25; Eidelberg, pp. 100–1 and 22; Chazan, pp. 227 and 244; Bernold of St Blaise, *Chronicon*, MGH, Scriptores, 34 vols (Hanover, 1826–1980) vol. 5, p. 465.
30. Neubauer and Stern, pp. 48–51 and 2; Habermann, pp. 95–7 and 25–6; Eidelberg, pp. 101–5 and 23; Chazan, pp. 228–32 and 245; Bernold of St Blaise, MGH, Scriptores, vol. 5, p. 465.
31. Neubauer and Stern, pp. 51–7 and 2–17; Habermann, pp. 97–104 and 26–43; Eidelberg, pp. 105–15 and 23–49; Chazan, pp. 232–42 and 245–73. Albert of Aachen, RHC Oc 4, pp. 292–3; the *Annalista Saxo*, MGH, Scriptores, vol. 6, p. 729; the *Annales Wirziburgenses*, MGH, Scriptores, vol. 2, p. 246.
32. Neubauer and Stern, pp. 17–25; Habermann, pp. 43–52; Eidelberg, pp. 49–61; Chazan, pp. 273–87. Albert of Aachen, RHC Oc 4, p. 292.
33. Neubauer and Stern, pp. 25–9; Habermann, pp. 52–6; Eidelberg, pp. 62–7; Chazan, pp. 287–93. *Gesta Treverorum*, MGH, Scriptores, vol. 8, p. 190.
34. Neubauer and Stern, p. 29; Habermann, p. 56; Eidelberg, p. 67; Chazan, p. 293.
35. Neubauer and Stern, p. 29; Habermann, p. 56; Eidelberg, p. 67; Chazan, p. 293.
36. Cosmos of Prague, *Chronica Boemorum*, MGH, Scriptores, vol. 9, p. 103; the *Annalista Saxo*, MGH, Scriptores, vol. 6, p. 729. In the *Solomon bar Simson Chronicle*, there is reference to yet one more locale, identified with the three Hebrew consonants *sh-l-'*. However, the story told about the Jews of *sh-l-'* is of questionable reliability, and I have chosen to omit it from this catalogue of more-or-less reliably reported events.
37. See the previous note on this unlikely story.
38. The role of Count Emicho in the assault on Mainz Jewry is highlighted in the *Mainz Anonymous*, as we have seen. The *Solomon bar Simson Chronicle* seconds this view, although I am inclined – as noted – to see its account of

events in Mainz as based on the earlier *Mainz Anonymous*. Albert of Aachen likewise associates the destruction of Mainz Jewry with Emicho.

39. Neubauer and Stern, pp. 18–19; Habermann, p. 45; Eidelberg, p. 51; Chazan, p. 275.
40. Neubauer and Stern, p. 20; Habermann, p. 46; Eidelberg, p. 53; Chazan, p. 278.
41. Neubauer and Stern, p. 21; Habermann, p. 48; Eidelberg, p. 55; Chazan, p. 280.
42. Neubauer and Stern, p. 23; Habermann, pp. 50–1; Eidelberg, pp. 58–9; Chazan, p. 284.
43. Albert of Aachen's report on the destruction of Cologne Jewry is brief and problematic. Like the *Solomon bar Simson Chronicle*, Albert attributes the initial violence in Cologne itself to the burghers, as we shall see more fully below. Beyond that violence, however, he knows only of assault on Cologne Jews seeking to make their way to Neuss, an attack that he attributes loosely to 'peregrini and cruce signati' (pilgrims and wearers of the cross).
44. Cosmos of Prague highlights the role of the crusaders; the *Annalista Saxo* does not tell us anything with respect to the identity of the attackers.
45. The only numerical estimates provided for Jewish casualties are found in the *Solomon bar Simson Chronicle*. The mid-twelfth-century author/editor suggests 800 Jewish deaths in Worms and 1100 in Mainz – see Neubauer and Stern, pp. 2 and 8; Habermann, pp. 25 and 32; Eidelberg, pp. 23 and 33; Chazan, pp. 245 and 256. I have tentatively suggested approximately the same number of casualties for Cologne as for Mainz, thus giving a sense of the heavy toll exacted by the crusader assaults on these two major Rhineland Jewish settlements.
46. Neubauer and Stern, p. 29; Habermann, p. 56; Eidelberg, p. 67; Chazan, p. 293.
47. I have analysed the Trier segment of the *Solomon bar Simson Chronicle* in *God, Humanity, and History*, chap. 4. For comparative observations on this Hebrew source and the *Gesta Treverorum* account of events in Trier, see E. Haverkamp, ' "Persecutio" und "Gezerah" in Trier während des Ersten Kreuzzugs', in A. Haverkamp (ed.), *Juden und Christen zur Zeit der Kreuzzüge*, Vorträge und Forschungen, 47 (Sigmaringen, 1999) pp. 35–71, and Robert Chazan, 'Christian and Jewish Perceptions of 1096: A Case Study of Trier', *Jewish History*, **13** (1999) 9–22. For our present purposes, I shall not cite the *Gesta Treverorum*, which focuses only in the most minimal fashion on the realties of the persecution.
48. Neubauer and Stern, pp. 17–18; Habermann, pp. 43–4; Eidelberg, pp. 49–50; Chazan, 274.
49. Albert of Aachen, RHC Oc. 4, p. 292.
50. Again, recall that it was Emicho and his troops who were responsible for the systematic destruction of the enclaves in which the Jews of Cologne had hoped to find refuge.
51. Neubauer and Stern, p. 25; Habermann, p. 53; Eidelberg, p. 62; Chazan, p. 288.
52. The crusader–burgher coalition indicated involved – we must assume – random groups of crusaders and townsmen. The organized forces associated with Peter the Hermit had long since departed eastward.
53. Neubauer and Stern, pp. 23–4; Habermann, pp. 51–2; Eidelberg, pp. 59–60; Chazan, pp. 284–6.
54. For some of the literature on papal thinking, see above, n. 1.

55. For Bernard's arguments against assault on Jews, see his crusade letter in J. Leclercq and H. M. Rochais (eds), *Sancti Bernardi opera*, 8 vols (Rome, 1957–77) vol. 8, p. 316. I have analysed these arguments in some detail in *Medieval Stereotypes and Modern Antisemitism* (Berkeley and Los Angeles, CA, 1997) chapter 3.

56. Again, see the works cited in n. 1.

57. The revenge motif in the 1096 assaults is emphasized in J. Riley-Smith, 'The First Crusade and the Persecution of the Jews', in: W. J. Sheils (ed.), *Persecution and Toleration*, Studies in Church History, vol. 21 (Oxford, 1984) pp. 51–72.

58. For broad analysis of these distortions, see my *European Jewry and the First Crusade*, pp. 65–81.

59. The failure of the Hebrew narratives to be sufficiently explicit with respect to the sources of burgher antipathy and violence may well be intentional, rooted in the sense that burgher hostility was less ideologically grounded and thus conferred less dignity on the Jewish victims.

60. See the discussion of burgher violence in Trier above, and the specific citation identified in n. 51.

61. Neubauer and Stern, p. 25; Habermann, p. 53; Eidelberg, p. 62; Chazan, p. 288.

62. We possess both Christian and Jewish sources for the founding of Speyer Jewry. See the charter of Bishop Rudiger, in A. Hilgard (ed.), *Urkunden zur Geschichte der Stadt Speyer* (Strasbourg, 1885) p. 11, and a Hebrew narrative account in Neubauer and Stern, p. 31; Habermann, pp. 59–60; Eidelberg, pp. 71–2.

63. Neubauer and Stern, pp. 50–1; Habermann, p. 97; Eidelberg, p. 105; Chazan, pp. 231–2.

3
Christian Theology and Anti-Jewish Violence in the Middle Ages: Connections and Disjunctions*

Jeremy Cohen

The linkage between theology – or the ideology of any regnant establishment or intellectual elite, for that matter – and the infliction of physical violence may often appear indirect and questionable, perhaps even tenuous and misleading. In the late ancient and medieval history of Christian–Jewish relations, counter-examples abound. On the one hand, in numerous instances of harshly anti-Jewish preaching – as in John Chrysostom's late fourth-century Antioch, Isidore of Seville and Julian of Toledo's seventh-century Visigothic Spain, and Lyons of the ninth-century Bishops Agobard and Amulo – little concrete evidence attests to increased physical attacks on Jews or an overall decline in Jewish well-being. On the other hand, notwithstanding the vociferous protestations of popes and emperors, late medieval Jewish communities frequently suffered serious losses of life and property as a result of libels of ritual murder, ritual cannibalism (blood libel) and host desecration. Moreover, as we learn from recent studies like David Nirenberg's *Communities of Violence*, the place of violent behaviour – sporadic, widespread or ritualised – in a given socio-cultural context is often multivalent, serving not only to harm and destroy its targets but perhaps even to stabilize and protect them.[1]

And yet, while one ought not blindly to postulate a direct cause-and-effect relationship between Christian theology and anti-Jewish violence in the Middle Ages, one cannot deny that Christian anti-Judaism took its toll in the history of the medieval Jewish experience, and I shall seek here to illustrate some aspects of the complicated process whereby it did. By way of example, I shall consider two manifestations of Christian violence during the high Middle Ages: that inflicted upon the persons of the Jews of northern Europe during the First and Second Crusades, and that directed against the books of the Jews during the 1240s.

In order to appreciate the complexity of the relationship between theology and violence in these instances, one must take note from the outset that Jews were not supposed to suffer violence in a properly ordered and maintained Christian society. The apostle Paul had prophesied the future conversion of all Israel to Christianity, leading many to infer that the Jewish presence in Christendom could not be violently eradicated. And, setting the tone for ecclesiastical policy throughout much of the Middle Ages, Augustine of Hippo elaborated much more specifically concerning the divine mandate for Jewish survival.

> For there is a prophecy given previously in the Psalms . . . : 'Slay them not, lest at any time they forget your law; scatter them by your might.' God thus demonstrated to the church the grace of his mercy upon his enemies the Jews, since, as the Apostle says, 'their offence is the salvation of the Gentiles.' Therefore, he did not kill them – that is, he did not make them cease living as Jews, although conquered and oppressed by the Romans – lest, having forgotten the law of God, they not be able to offer testimony on our behalf. Thus it was inadequate for him to say, 'slay them not, lest at any time they forget your law', without adding further, 'scatter them'. For if they were not everywhere, but solely in their own land with this testimony of the scriptures, the church, which is everywhere, could not have them among all the nations as witnesses to the prophecies given previously regarding Christ.[2]

Aside from testimonial value of the Jew rationalizing Augustine's call for his survival, this passage has important practical applications: the dispersion of the Jews throughout all of Christendom is a desideratum facilitated by the loss of their homeland and their enforced subjugation by Christian rulers ('scatter them by your might' – elsewhere he speaks of servitude). Nevertheless, one may not kill the Jews. Equally important, the injunction to 'slay them not' extends beyond the prohibition of physical violence to a ban on interference with the practice of Judaism ('he did not kill them – that is, he did not make them cease living as Jews'). During most of the Middle Ages, popes, prelates and theologians followed the Augustinian lead. Gregory the Great, for instance, sharply chastized Christians who sought to baptize Jews against their will, to harm them and their property, and to obstruct their observance of Jewish rites. With words that eventually became a formulaic guarantee of their right to live in Christendom free of violent attack (*Sicut Iudaeis non . . .*), Gregory instructed: 'Just as the Jews should not have license in their synagogues to arrogate anything beyond that permitted by law, so too in those things granted them they should experience no infringement of their rights.' Even Pope Innocent III, whom no historian ever considered a friend of the Jews, reaffirmed the Gregorian rule of *Sicut Iudaeis non* some six centuries later and invoked the legacy of Augustine as well, noting that 'they ought not be killed, lest the Christian people forget the divine law'.[3] Two generations after Innocent, Thomas Aquinas concluded:

> Though God is omnipotent and supremely good, he permits certain bad things, which he could prevent, to transpire in the universe, lest, without them, greater goods might be forfeited or worse evils ensue. . . . From the fact that the Jews observe their rituals, in which the truth of the faith we now hold was once prefigured, there proceeds the good that our enemies bear witness to our faith, and that which we believe is represented in a figure.[4]

Not only did this logic protect the persons of the Jews and their practice of Judaism, but Thomas encouraged respect for Jewish property beyond what the law demanded. If, owing to the Jews' perpetual servitude, Christian princes can by rights 'treat their property as if it were their own, save for this restriction – that the subsistence necessary for life is not taken away from them', Thomas urged that, in practice, rulers not exceed established custom in taxing their Jewish subjects.[5]

With these considerations in mind, let us move on to our two case studies. At the end of the spring and beginning of the summer of 1096 – during the earliest months of the First Crusade – bands of armed crusaders attacked Jewish settlements in western and central Germany; those Jews whom they could the crusaders converted, while others who fell in their path they killed. Jewish communities of the Rhine valley – in Speyer, Worms, Mainz, Cologne and its suburbs, Metz and Trier – and others including Regensburg and Prague to the east, suffered serious losses in life and property upon this, the first widespread outbreak of anti-Jewish violence in medieval Christian Europe.[6]

Why did the massacres occur? Some have linked the pogroms to economically grounded hostility and jealousies that characterized the relations between German Jews and their neighbours, or, perhaps, to factionalism that dominated German politics in the wake of the Investiture Conflict. Yet the preponderance of the evidence from both Jewish and Christian sources indicates that religious zeal motivated the attackers above all. Consider the rationale attributed to the crusaders in a Jewish chronicle of the First Crusade, written in Hebrew during the first half of the twelfth century:

> Why should we concern ourselves with going to war against the Ishmaelites dwelling about Jerusalem when in our midst is a people who disrespect our God – indeed, their ancestors are those who crucified him. Why should we let them live and tolerate their dwelling among us? Let us use our swords against them first and then proceed upon our 'stray' path.[7]

While the description of the crusade as a 'stray path' clearly reflects caricaturing by the Jewish chronicler, this rationale for attacking the Jews finds confirmation in Guibert of Nogent's description of an attack on the Jews of Rouen in 1096:

> the people who had undertaken to go on that expedition under the badge of the cross began to complain to one another, 'After traversing great distances, we desire to attack the enemies of God in the East, although the Jews, of all races the worst foes of God, are before our eyes. That's doing our work backward.'[8]

Such conviction also found expression in Raymond of Aguilers' Latin chronicle of the First Crusade, which recounts how God spoke of the Jews to the crusaders, as they neared the Holy Land in 1099: 'I entertain hatred against them as unbelievers and rank them the lowest of all races. Therefore, be sure you are not unbelievers, or else you will be with the Jews, and I shall choose other people and carry to fulfilment for them My promises which I made to you.'[9] After numerous, intense battles against the Saracens, the Jews still exemplified religious infidelity; they, even more than the Muslims, were the enemies of God *par excellence*, the living antithesis to God's covenant with his chosen people.

Nevertheless, the crusaders had no official licence to attack the Jews, and the relative ease with which forcibly converted Jews openly returned to Judaism in the aftermath of the First Crusade suggests that the Christian establishment acknowledged the illegitimacy of the violence. Early in the 1060s – in his subsequently canonized bull, *Dispar nimirum est* – Pope Alexander II had specifically admonished the Christian warriors of the Spanish Reconquista that 'it is impious to wish to annihilate those who are protected by the mercy of God' and that 'the situation of the

Jews is surely different from that of the Saracens'.[10] How is it, then, that theological/ecclesiastical strictures against anti-Jewish violence fell on deaf ears? Many scholars, including some of the authors in this volume, have struggled with these questions. In a climate of intense religious zeal and desire for vengeance, perhaps appreciating the balance between the negative ramifications of *Adverus Judaeos* teaching and the need to refrain from attacking the Jew exceeded the grasp of many a common crusader. Had he taken a more explicit stance in 1095–6 when preaching the crusade, perhaps Pope Urban II could have stifled some of the rhetoric and enthusiasm that resulted in the pogroms.

Whether or not one accepts these suppositions, I believe that one must appreciate the complexity of the construction of the Jew at the heart of medieval Christian thought. In another context, I have argued that Bernard of Clairvaux, the famed Cistercian abbot who went to great lengths to protect the Jews against physical harm during the Second Crusade, actually contributed to a mentality that precipitated anti-Jewish violence in the very letters in which he warned against it. Bernard construed Muslims and Jews as the embodiments of sinful evils plaguing Christian society; the crusade, in turn, comprised a divinely ordained opportunity for Christians to direct their passions towards commendable ends and thereby to purge Christendom of the turpitude impeding its salvation. Granted that Bernard heartily affirmed the Augustinian maxim of 'slay them not', his Jews represented an evil that God's faithful had to overcome and excise from within their midst; and the crusade, Bernard maintained, constituted the perfect opportunity for doing so.[11] Alongside Bernard, I would here propose considering his Benedictine contemporary, Peter the Venerable of Cluny; in the same year (1146) that Bernard issued his letters calling for participation in the Second Crusade and yet forbidding physical harm to the Jews, Peter wrote to King Louis VII of France, applauding his leading role in the crusade and addressing the matter of the Jews in particular.

Peter's Letter 130[12] first expresses unmitigated support for the crusade, despite the abbot's inability to participate personally in the expedition: Christendom has witnessed a renewal of the ancient miracles whereby Moses led the Israelites out of Egypt and Joshua led them in to the promised land; surely the merits of a Christian king like Louis must outweigh those of the Jews, who themselves enjoyed the providence of God in their conquests. Second, the letter turns to the Jews of Peter's own day in light of the struggle against the Saracens. Echoing the reasoning that impelled crusaders to attack the Jews of northern Europe in 1096, as understood by both Jewish and Christian chroniclers of the crusade, Peter wondered out loud, as it were, concerning the rationale underlying the call for the new crusade.

> Towards what end should we pursue and persecute the enemies of the Christian faith in far and distant lands if the Jews, vile blasphemers and far worse than the Saracens, not far away from us but right in our midst, blaspheme, abuse, and trample on Christ and the Christian sacraments so freely and insolently and with impunity? How can zeal for God nourish God's children if the Jews, enemies of the supreme Christ and of the Christians, remain totally unpunished?

To be sure, Peter quickly tempered his outcry with the qualification that he did not wish for Christians to attack the Jews physically.

> I recall that written about them in the divine psalm, the prophet speaking thus in the spirit of God: 'God', he said, 'has shown me in the matter of my enemies, slay them not.' For God does not wish them to be entirely killed and altogether wiped out, but to be preserved for greater torment and reproach, like the fratricide Cain, in a life worse than death.... So the fully just severity of God has dealt with the damned, damnable Jews from the very time of the passion and death of Christ – and will do so until the end of time. Those who shed the blood of Christ, their brother according to the flesh, are enslaved, wretched, fearful, mournful, and exiled on the face of the earth – until the remnants of this wretched people shall turn to God once, as the prophet has taught, the multitude of the Gentiles has already been called.

Third, in practical terms, how ought Christian society to enforce this servile status of Jews upon them? Peter called for depriving the Jews of much of their wealth, which they have not acquired honestly by working the land (*de simplici agri cultura*) but rather by deceiving Christians. Compounding their blasphemy, the Jews traffic in stolen property and, most offensively, in relics, icons and other ritual objects stolen from churches and then sold to the synagogues of Satan, where Christian thieves find refuge as well. Fourth, and finally, the letter returns to the pressing issue of the day, the crusade: To the extent possible, one should levy the financial cost of the holy war upon the Jews: 'Let their lives be spared but their money taken away', rather than imposing the entire burden of the crusade upon the Christian faithful.

Like Bernard, in speaking out in support of the crusade Peter presented guidelines for combating the infidels who endanger the security of

Christendom, and thus did he come to address the nature of the Jews alongside that of the Muslims. Peter, too, defined the Jewish threat to Christian society largely in economic terms; and he linked the economic activity of the Jews to their allegedly characteristic carnality on one hand, while comparing it to that of Christian thieves on the other hand. Peter echoed his Cistercian colleague's sympathy with the zeal motivating crusaders to attack the Jews, by espousing their very rationale for such violence. Yet Peter also proceeded to oppose violent attacks upon the Jews, citing the Augustinian proof-text of Psalm 59: 12, while at the same time seeking to curb the Jews' economic exploitation of Christian society. Even more emphatically than Bernard, Peter advocated that the Jews somehow share in the financial burden of the crusade.

No less important, just as Bernard addressed the matter of the Jews from the perspective of his crusading theology, so can one better understand Peter's letter through his appraisal of the Christian campaigns to liberate the Holy Land from Muslim control. Beyond its endorsement of the crusade, likening it to the divinely ordained wars of the ancient Israelites, the letter to Louis VII affords little insight into Peter's crusading ideology; but his sermon 'In Praise of the Lord's Grave' (*De laude dominici sepulchri*), preached in the presence of the pope in the same year or the next, offers more.[13] There Peter extolled the sanctity and symbolic importance of the Holy Sepulchre at length. Jesus' grave connotes the chief reason for exulting in Christ, that which truly facilitates Christian victory over the enemies of God. The death of Jesus, his resurrection and his ascension to heaven outweigh his birth in their importance, and the honour of the sepulchre accordingly exceeds that of the manger in Bethlehem. Scripture appropriately attests to the centrality of the Holy Land and of Jerusalem in a Christian view of the world. Just as Jesus' grave at the heart of the earth contained his body, so ought the Christian to embody the eternal memory of Christ in his heart, centrally located among human organs. Today the sepulchre embodies the Christian hope for final salvation, hope which has now spread throughout the entire world, only a few remaining Jews and the wicked sect of Muhammad excepted. God has confirmed this status of the sepulchre and the hope that proceeds therefrom in numerous ways, above all in the miraculous fire that kindles the lamps in the Holy Sepulchre every Easter – a miracle whose virtues and veracity the sermon painstakingly elaborates. Here, then, lies the route to salvation: forsaking the pleasures of this world, a Christian must dedicate himself to the holiness, memories and miracles enshrined in the grave of his saviour, joining the universal convocation of faithful souls that it has attracted, liberating it from the baseness of the infidels.

Moreover, like Bernard, Peter construed the Jews and their lifestyle as exemplifying the antithesis of that which the crusader should emulate. The letter to Louis VII underscores his association of the Jews with the sinful, misguided pursuit of financial profit. And from Peter's perspective on salvation history, the *De laude* develops the opposition further: the Jews do not interpret Scripture properly, so as to fathom the grandeur of Jerusalem or its holy sites and discern the way towards eternal life. The Jews murdered the body of Christ that gave the sepulchre its sanctity. The annual miracle of the fire in the Holy Sepulchre, recalling the flame with which God accepted Abel's sacrifice over Cain's and Elijah's over that of the prophets of Baal, carries yet an additional message of contemporary relevance.

> And so at the present time, O Lord...do you clearly distinguish between us and the Jews or pagans; thus do you spurn their vows, their prayers, and their offerings; thus do you show that these are repugnant to you. Now that their offerings have been rejected, you approve of ours. In this way do you proclaim that the sacrifices, prayers, and vows of your Christians are pleasing to you: You direct a fire to proceed from heaven to the grave of your son, which only they respect and revere; with that same fire you set their hearts on fire with love for you; with its splendour do you enlighten them, now and forever. And since the perfidious enemies of your Christ disparage his death more than his other acts of humility, in adorning the monument of his death with a miracle of such light do you demonstrate how great is the darkness of error in which they are confined. While they despise his death above all, you honour the monument of his death above all; what they consider particularly shameful you prove to be especially glorious by means of so wonderful a sign. You reject the Jews like the hateful Cain, the pagans like the worshippers of Baal, and you do not light a fire on their offerings. Yet you do desire the hosts of the Christian people, just like the offerings of Abel; you approve of its sacrifice, like the holocaust of Elijah, and thus with a fire sent from heaven do you irradiate the grave in which your son, offered as a sacrifice on our behalf, lay at rest.[14]

The miracle of fire at Jesus' grave attests to the rejection of the synagogue, on the model of Cain, and the Saracens, on the model of the idolatrous prophets of Baal. Jews and Saracens in Peter's eyes together epitomized the threat of *infidelitas*, which endangered Christendom and which therefore, paradoxically, held out the promise of salvation for those who would join the crusade.

Peter's crusading ideology, as expressed in his *De laude*, can lead us to an enhanced appreciation of his epistle to Louis VII. Curiously, this text displays a chiastic structure, which unfolds in a series of four *a fortiori* arguments, corresponding to the four central points of the letter as outlined in our summary:

(i/a) *The Jews of old vs. the crusader king (Louis VII)*:   If God ensured the victories of the former, how much more should he assist the latter. For '...the former observed the divine commandments, but, to a certain extent, they exerted themselves in combat out of hope for an earthly reward; yet the latter endangers and even sacrifices his kingdom, his wealth, and even his life...so that, after the disappearance of his mortal kingdom, he might be crowned with honour and glory by the king of kings.'

(ii/b) *Saracens vs. Jews*:   If the Saracens are detestable because, although they acknowledge (as we do) that Christ was born of a virgin and they share many beliefs about him with us, they reject God and the son of God (which is more important) and they do not believe in his death and resurrection..., how much more must we curse and hate the Jews who, believing nothing concerning Christ or the Christian faith and denying the virgin birth and all the sacraments of human salvation, blaspheme and insult him?'

(iii/b') *Christian thieves vs. Jews*:   While the former suffer capital punishment for trading in stolen church property, the latter go unpunished, owing to that ancient, satanic law that protects them. 'The Jew grows fat and revels in his pleasures, while the Christian hangs from a noose!' Surely, Peter implied, the opposite should hold true.

(iv/a') *The Jews of old vs. the crusaders*:   'Just as once, when the ancestors of the Jews were still in God's favour, the riches of the Egyptians were given over to their possession according to divine command', how much the more so, one reasons, 'should the wealth of the Jews, even against their will, serve the needs of Christian peoples.'

The outer poles of the chiasmus compare the victories of the ancient Israelites with those of King Louis and the crusaders: if God saved and rewarded the Jews when they still found favour in his eyes, even though a craving for earthly profit rendered their faith imperfect, he certainly should ensure the military success and financial feasibility of the crusade.

For their part, the inner vertices of the chiasmus contrast the Jews with both external and internal enemies of the church: owing to their greed and their blasphemy, the Jews are more injurious than others; if Christendom justly inflicts punishment on Saracens and Christian thieves, surely the Jews ought to suffer too. The structure of Peter's letter evidently confirms the thrust of its contents. Throughout their history, ancient and modern, the Jews have embodied greed and *infidelitas*. Having inherited the spiritual election of the ancient synagogue (a), the church now receives God's help to crusade against the forces of blasphemy and carnality both outside Christendom (b) and within (b') – in either case exemplified by the Jew – and, on the model of ancient Israel, it may appropriate the assets of its enemies toward this end (a'). Peter, like Bernard, construed the Jews and Judaism from his particular mid-twelfth-century vantage point, at which crusading, fraught with ecclesiological and eschatological significance, helped to crystallize the world-view of a Christian theologian. One can readily discern how those who dedicated themselves to armed combat in the name of the cross might construe the tone and substance of Peter's declarations – in the documents considered here and elsewhere – a mandate for violence against the Jews, even if Peter himself never enunciated that conclusion.

We recall that Augustine's injunction against slaying the Jews extended beyond physical injury to interference with their observance of their religious law. In medieval Christian terms, then, anti-Jewish violence included violence against the maintenance of Jewish religious life; and, for our second example, we turn to the persecution of the Talmud by the late medieval Church, commencing in the 1230s and 1240s. The story of this persecution has been well told, retold and debated in recent research,[15] and our present goal is to see how it, too, reveals basic ambivalence in Christian theological perceptions of the Jew: one who commands a legitimate place in a properly ordered Christian society, but one who exemplifies the evil subverting the integrity of that society, whose eradication might draw that society closer to perfection.

To summarize briefly: Pope Gregory IX heard the Talmud denounced by Nicholas Donin, an embittered Jewish apostate, in 1236. Three years later, Pope Gregory issued a series of condemnatory bulls, ordering rulers and prelates of Christian Europe to impound the Talmud and other Jewish writings on the first Sabbath during Lent in 1240, and to submit the books to ecclesiastical authorities for inspection. For whatever reason, Gregory's decrees were implemented only in the royal domains of France. There King Louis IX confiscated rabbinic texts, summoned leading French rabbis to his court in 1240 to defend the Talmud against

They prohibit children from using the Bible because, as they say, it is not fit for instruction, but, preferring the doctrine of the Talmud, they have given various commandments of their own accord.[16]

When Gregory IX issued his letters condemning the Talmud, he wrote of the Jews of France and of other lands that,

> not content with the Old Law which God gave in writing through Moses, and even ignoring it completely, they affirm that God gave them another Law which is called the Talmud, that is teaching. They lie to the effect that it was handed down to Moses orally and implanted in their minds, and was preserved unwritten for a long time until there arrived those whom they call sages and scribes, who reduced it to writing so that it not be forgotten from people's minds. Its written version exceeds the text of the Bible in size. In it are contained so many abusive and wicked things, that they are an embarrassment for those who mention them and a horror for those who hear them.[17]

One finds Donin's influence on Gregory unmistakable, but this hardly deprives the pope of responsibility for the contents of his own decree. A marked sensitivity to doctrinal irregularity comports well with the overriding concerns and interests of Gregory IX, who sought to oversee the curriculum at the University of Paris, to prohibit the study of the unexpurgated works of Aristotle, to promote legislative uniformity in Christendom through the promulgation of his *Decretales* (Decretals), and to intensify the campaign of his church against heresy. Gregory, in fact, exceeded Donin's 35 accusations in claiming explicitly that the Jews have substituted another law (*legem aliam*) for that of the Bible.

The French rabbis compelled to defend the Talmud in debate with Nicholas Donin opened and rested their case with a lengthy defence of the Talmud's antiquity and inextricable ties to the divinely revealed law of Moses: 'The Talmud is an explanation [of Scripture], and were it not for the Talmud a person could not understand the commandments thoroughly.'[18] More important, as Chen Merchavia's careful analysis has established,[19] the proceedings in Paris in 1240 included not only the disputation between Donin and the French rabbis, but also the more formal ecclesiastical inquiry. This official proceeding seems not to have included Donin and is not recorded in any Hebrew source; it involved a number of distinguished prelates and theologians, who reviewed the evidence of the confiscated Jewish books and interrogated the rabbis. The most decisive appraisal of the clerical ruling that the Talmud should burn comes from the pen of one of these ecclesiastical investigators, Odo of Châteauroux, who then served as chancellor of the University of Paris. Odo, we recall, was later enlisted by Pope Innocent IV to head another formal inspection of the Talmud, and, upon receipt of that commission in 1247, he wrote to Innocent to summarize the ecclesiastical actions undertaken thus far. After reproducing the texts of Pope Gregory's decrees, he recounted the findings of the previous Parisian board of inquiry:

> It was discovered that the said books were full of errors, and that a veil has been placed over the heart of these people to such an extent that they [the books] turn the Jews away from not only a spiritual understanding [of the law] but even from a literal understanding, and they incline them to fantasies and lies. . . . After the said examination had been made, and the advice of all the teachers of theology and canon law, and of many others, had been taken, all the said books which could be secured were incinerated by fire in accordance with the apostolic decree.[20]

Heading the ecclesiastical investigators' verdict that ordered the Talmud to be burned stood the charge that not only did it deter the Jews from Christianity, but it steered them away from the literal observance of Mosaic law. Talmudic Judaism did not hold true to the biblical faith and observance whose toleration Augustine had repeatedly preached; by implication, the Talmudic Jew did not serve the purpose that justified the Jewish presence in Christendom. Accordingly, when Pope Innocent IV renewed his predecessor's condemnation of the Talmud in 1244, the Jewish rabbis of France appealed to him (as the pope later recounted) that the opposite was the case, 'that without that book which in Hebrew is called "Talmut" they cannot understand the Bible and the other statutes of their law according to their faith'.[21] Again charged with investigating the Talmud and the charges against it, Bishop Odo now reported to the pope that 'the masters of the Jews of the kingdom of France recently lied to your holiness . . . saying that without those books, which are called the Talmud in Hebrew, they cannot understand the Bible and the other provisions of their law in keeping with their faith'.[22]

If Augustine had proscribed violence against Judaism because it bore witness to the biblical foundations of Christianity, the thirteenth-century church now explained its attack on contemporary Judaism by denying that its proponents performed the function that rationalized their preservation. Echoed by popes who followed him, Gregory IX contended that the Talmud bears primary responsibility for the Jewish refusal to accept

Christianity; he termed it 'the primary cause which holds the Jews obstinate in their perfidy'.[23] As several historians have noted, Pope Innocent IV moved readily from voicing his rationale for policing the doctrinal beliefs of infidels to advocating aggressive missionizing amongst them.[24] The attack on rabbinic Judaism facilitated and then accompanied unprecedented efforts to convert the Jews of Latin Christendom, an undertaking which likewise strayed from the legacy and logic of Augustinian doctrine.[25]

Nonetheless, the most impressive qualities of the attitude towards the Jew and his books in late medieval theology remain its ambivalence and hesitation. When the Jews of France protested his condemnation of the Talmud, Innocent IV did not dismiss their appeal summarily; instead, he wrote to the king of France: 'We who, in keeping with the divine injunction, are obligated to sustain them in the observance of that law therefore considered it proper to respond to them thus: We do not want to deprive them of their books unjustly, if in so doing we should deprive them of the observance of their law.'[26] Just two or three months earlier, Innocent had sought to defend Jews charged with ritual murder against the violence of those acting 'contrary to the clemency of the Catholic religion which allows them to dwell in the midst of its people and has decreed tolerance for their rites', inasmuch as they 'had been left as witnesses of his saving passion and his victorious death'.[27]

Some have argued that Innocent, faithful to the Augustinian legacy, in fact retreated from Gregory IX's outright condemnation of the Talmud as heretical and laid the groundwork for those who called for Jewish books to be censored of their specifically abusive passages and then returned to the Jews.[28] My own sense is somewhat different. As we have seen, the notion that, owing to its deviation from ancient biblical religion, the Talmud did not qualify for protection from Christian violence pervades the sources attesting to its condemnations during the pontificates of Gregory IX and Innocent IV. In his commentary on the Decretals, Innocent IV himself explained his own actions and Gregory's in such terms, and later medieval canonists and inquisitors followed suit.[29] Subsequent ecclesiastical actions against the Talmud attacked both its heretical character and its offensive passages, and one might well justify the inspection – rather than wholesale burning – of Jewish books with charges of Talmudic heresy, as Pope Clement IV did in 1267.[30] Inasmuch as Christian missionaries had begun to quote from the Talmud in preaching to the Jews, censorship (as opposed to destruction) might also comport well with new conversionist efforts among them, which themselves suggested that the Jewish presence was no longer the desideratum perceived by Augustine.

I would conclude that the stance of the late medieval papacy *vis-à-vis* the Talmud defies tidy generalization. After the middle of the thirteenth century, most medieval popes displayed little or no interest in rabbinic Judaism and its post-biblical literature; no doubt Jews of the time deemed that alternative the best of all. Those who did take a concerted stance undoubtedly proceeded from a variety of motivations. They determined the particulars of their policies out of consideration for conflicting interest groups, including the Jews themselves, for the better welfare of the Church and the papacy in its numerous dimensions, and for the precepts of Christian doctrine. One ought not to assume that theological considerations necessarily took precedence in such formulation of papal policy; instead, one should be quite surprised if they always did. But even when they did, those theological considerations were seldom univocal. The same churchman who violently persecuted the Talmud on theological grounds might also guarantee the physical safety of the Jews on theological grounds. As we have seen, ideologies can precipitate violent acts even when they do not openly condone them.

Notes

* Grateful acknowledgement is made to the University of California Press for permission to republish material that has appeared recently in my *Living Letters of the Law: Ideas of the Jew in Medieval Christianity* (Berkeley, Los Angeles and London, CA, 1999).

1. See D. Nirenberg, *Communities of Violence: Persecution of Minorities in the Middle Ages* (Princeton, NJ, 1996).
2. Augustine, *De Civitate Dei*, 18.46, ed. B. Dombart and A. Kalb, CCSL 48 (Turnhout, 1955) pp. 644–5.
3. Gregory, *Registrum Epistularum*, 8.25, ed. D. Norberg, CCSL 140A (Turnhout, 1982) pp. 546–7. On the subsequent history of the *Sicut Iudeis* bull and formula, see S. Grayzel, 'The Papal Bull *Sicut Iudeis*', in M. Ben-Horin et al. (eds), *Studies and Essays in Honor of Abraham A. Neuman* (Leiden, 1962) pp. 243–80; and S. Simonsohn, *The Apostolic See and the Jews: History*, Pontifical Institute of Mediaeval Studies: Studies and Texts, 109 (Toronto, 1991) pp. 39–93.
4. Thomas Aquinas, *Summa theologiae*, 2a2ae, qu. 10, art. 11, ed. and trans. Blackfriars edn, vol. 32 (London, 1975) pp. 72–3 (with modifications to the Blackfriars translation).
5. Thomas Aquinas, *Epistola ad ducissam Brabantiae*, ed. R. Busa, in *Opera omnia*, vol. 3 (Stutttgart, 1980) pp. 594–5.
6. On the anti-Jewish violence of 1096, see above all R. Chazan, *European Jewry and the First Crusade* (Berkeley and Los Angeles, CA, 1987); on the appreciation

of 1096 as a major turning-point in medieval Jewish history, see also S. Schwarzfuchs, 'The Place of the Crusades in Jewish History' [Hebrew], in M. Ben-Sasson et al. (eds), *Culture and Society in Medieval Jewry: Studies Dedicated to the Memory of Haim Hillel Ben-Sasson* (Jerusalem, 1989) pp. 251–69; J. Cohen, 'Recent Historiography on the Medieval Church and the Decline of European Jewry', in J. Ross Sweeney and S. Chodorow (eds), *Popes, Teachers, and Canon Law in the Middle Ages: Essays in Honor of Brian Tierney* (Ithaca, NY, 1989) pp. 251–62; and A. Grossman, 'Shorashav shel Kiddush ha-Shem b^e-Ashk^enaz he-Q^edumah', in I. M. Gafni and A. Ravitzky (eds), *Sanctity in Life and Martyrdom: Studies in Memory of Amir Yekutiel* (Jerusalem, 1992) pp. 100–5.

7. A. Neubauer and M. Stern (eds), *Hebräische Berichte über die Judenverfolgungen während der Kreuzzüge*, Quellen der Geschichte der Juden in Deutschland, vol. 2 (Berlin, 1892) p. 4, trans. and ed. S. Eidelberg, *The Jews and the Crusaders: The Hebrew Chronicles of the First and Second Crusades* (Madison, WI, 1977) p. 26.

8. Guibert of Nogent, *De vita sua sive Monodiae*, 2.5, ed. and trans. E.-R. Labande, *Les Classiques de l'Histoire de France au Moyen Age*, vol. 34 (Paris, 1981) pp. 246–8; trans. in J. F. Benton (ed.), *Self and Society in Medieval France: The Memoirs of Abbot Guibert of Nogent* (1970; repr. Toronto, 1984) pp. 134–5.

9. Raymond d'Aguilers, *Historia Francorum qui ceperunt Iherusalem*, trans. J. H. Hill and L. I. Hill, Memoirs of the American Philosophical Society, vol. 71 (Philadelphia, PA, 1968) p. 95.

10. S. Simonsohn, *The Apostolic See and the Jews: Documents, 492–1404*, Pontifical Institute of Mediaeval Studies: Studies and Texts, 94 (Toronto, 1988) pp. 35–6; trans. In R. Chazan (ed.), *Church, State, and Jew in the Middle Ages* (New York, 1980) p. 100.

11. J. Cohen, '"Witnesses of our Redemption": The Jews in the Crusading Theology of Bernard of Clairvaux', in B.-S. Albert et al. (eds), *Medieval Studies in Honour of Avrom Saltman*, Bar-Ilan Studies in History, vol. 4 (Ramat-Gan, Israel, 1995) pp. 67–81.

12. Peter the Venerable, *Letters*, ed. G. Constable, 2 vols (Cambridge, MA, 1967) vol. 1, pp. 327–30.

13. Peter the Venerable, *Sermones tres*, ed. G. Constable, *Revue Bénédictine*, **64** (1954) 232–54.

14. Ibid., p. 252. On the Christian monopoly on genuine sacrifice, see Peter the Venerable, *Contra Petrobrusianos*, 162, ed. J. Fearns, CCCM 10 (Turnhout, 1968) p. 95.

15. Among others, see J. Cohen, *The Friars and the Jews: The Evolution of Medieval Anti-Judaism* (Ithaca, N.Y., 1982) pp. 51–99; Ch. Merchavia, *The Church versus Talmudic and Midrashic Literature, 500–1248* [Hebrew] (Jerusalem, 1970), parts 2–3); S. Grayzel, 'The Talmud and the Medieval Papacy', in W. Jacob et al. (eds), *Essays in Honor of Solomon B. Freehof* (Pittsburg, 1964) pp. 220–45; K. R. Stow, 'The Burning of the Talmud in 1553, in Light of Sixteenth-century Catholic Attitudes toward the Talmud', *Bibliothèque d'Humanisme et Renaissance*, **34** (1972) 435–59; J. E. Rembaum, 'The Talmud and the Popes: Reflections on the Talmus Trials of the 1240s', *Viator*, **13** (1982) 203–23; R. Chazan, 'The Condemnation of the Talmud Reconsidered (1239–1248)', *Proceedings of the American Academy for Jewish Research*, **55** (1988) 11–30;

W. C. Jordan, 'Marian Devotion and the Talmud Trial of 1240', in B. Lewis and F. Niewöhner (eds), *Religionsgespräche im Mittelalter*, Wolfenbütteler Mittelalter-Studien, 4 (Wiesbaden, 1992) pp. 61–76; and Simonsohn, *The Apostolic See and the Jews: History*, pp. 300–42 – and the additional bibliography cited therein.

16. I. Loeb, *La Controverse sur le Talmud sous Saint Louis* (Paris, 1881) pp. 21–54. See also J. M. Rosenthal, 'The Talmud on Trial', *JQR*, n.s., **47** (1956) 58–76, 145–69; and Merchavia, *The Church*, Chapter 12.

17. Simonsohn, *The Apostolic See and the Jews: Documents*, p. 172; I have drawn from the translation in S. Grayzel (ed.), *The Church and the Jews in the Thirteenth Century*, revised edn (New York, 1966) p. 241.

18. *Wikkuah Rabbenu Yᵉhiel mi-Paris*, ed. S. Grünbaum (Thorn, 1873) pp. 2, 15–16.

19. Merchavia, *The Church*, pp. 240ff.

20. Grayzel, *The Church*, pp. 276–8, n. 3 (with departures from Grayzel's translation).

21. Simonsohn, *The Apostolic See and the Jews: Documents*, p. 197; Grayzel, *The Church*, pp. 274–5.

22. Grayzel, *The Church*, pp. 276–8, n. 3 (with departures from Grayzel's translation).

23. Simonsohn, *The Apostolic See and the Jews: Documents*, p. 172.

24. J. Muldoon, *Popes, Lawyers, and Infidels: The Church and the Non-Christian World, 1250–1550* (Philadelphia, PA, 1979) pp. 11ff; B. Z. Kedar, *Crusade and Mission: European Approaches Toward the Muslim* (Princeton, NJ, 1984) pp. 159ff; see also A. Melloni, *Innocenzo IV: La Concezione e l'esperienza della cristianità come regimen unius personae* (Genoa, 1990) esp. pp. 187–96, who places Innocent's outlook on Jews and Judaism within the broader context of his stance towards unbelievers and dissidents in general.

25. See Cohen, *The Friars, passim*.

26. Simonsohn, *The Apostolic See and the Jews: Documents*, p. 197; Grayzel, *The Church*, pp. 274–5.

27. Simonsohn, *The Apostolic See and the Jews: Documents*, pp. 190–2; Grayzel, *The Church*, pp. 262–5.

28. See J. E. Rembaum, 'The Talmud and the Popes', pp. 215ff; R. I. Burns, 'Anti-Semitism and Anti-Judaism in Christian History: A Revisionist Thesis', *Catholic Historical Review*, **70** (1984) 92–3; Simonsohn, *The Apostolic See and the Jews: History*, pp. 304–5; and K. R. Stow, *The '1007 Anonymous' and Papal Sovereignty: Jewish Perceptions of the Papacy and Papal Policy in the High Middle Ages*, Hebrew Union College annual, supplements, vol. 4 (Cincinnati, OH 1984) pp. 37–42 and *Alienated Minority: The Jews of Medieval Latin Europe* (Cambridge, MA, 1992) pp. 256ff.

29. B. Z. Kedar, 'Canon Law and the Burning of the Talmud', *Bulletin of Medieval Canon Law*, n.s., **9** (1979) 80–1; and cf. A. Funkenstein, *Perceptions of Jewish History* (Berkeley, CA, 1993) p. 195: 'The collection of materials for the trial, as well as the letters of Gregory IX and Innocent IV, show clearly that the "authority" of the Talmud, replacing that of the Bible, as well as its function as a "new law" (*nova lex*) were the primary concerns then.'

30. See the sources cited in Cohen, *The Friars*, pp. 78–81.

4

The Intellectual and Spiritual Quest for Christ and Central Medieval Persecution of Jews

Anna Sapir Abulafia

The central Middle Ages, broadly covering the period between 1050 and 1300, conjure up different images for different kinds of medieval historians. For crusade historians this period represents both the high point of crusading history – the conquest of Jerusalem in 1099 – and bitter disillusionment culminating in the fall of Acre in 1291. Socio-economic historians are offered a period of impressive growth in trade, industry and urbanization. Political and constitutional historians are struck by the burgeoning polities of the period. As intellectual historians feast themselves on the advancements in scholarship, ecclesiastical historians enjoy the fruits of reform. But however much students of Jewish history appreciate the intellectual, spiritual and economic growth of the Jewish communities of Ashkenaz (Germany and neighbouring lands) and Sefarad (Spain and neighbouring lands), they cannot overlook the reality of increasing pressure on, if not persecution of, Jews in this period. The obvious question to pose is what connections, if any, might exist between the many diverse facets of central medieval renewal and the hardening of Christian attitudes towards Jews, which commenced to the north of the Alps. In the context of this conference, commemorating the 1099 capture of Jerusalem by *milites Christi* (soldiers of Christ), it is tempting to examine the importance of the figure of Christ for the history of central medieval persecution of Jews. The link between the crusading ideal of wreaking vengeance on the enemies of Christ and the violence perpetrated against Jews during the crusades has been analysed most successfully by Jonathan Riley-Smith.[1] However, I want to go beyond that. If we take Bernard McGinn's definition of a crusade as 'a type of holy war that can be best understood as the foreign policy of the reform papacy for the reintegration of the Christian Roman Empire',[2] it makes sense to look more carefully at the role that the figure of Christ played in the reform movement of the period. For all their different approaches to reform, most reformers expressed their love for Jesus Christ. Spiritual reformers were filled with a longing for God and an intense interest in conversion of self and, in many cases, of the whole of society. But the quest for Christ was not just a spiritual affair. For twelfth-century scholars the question of *Cur Deus-homo* (Why God-man?) presented itself as a veritable intellectual challenge. Throughout the twelfth century monastic and scholastic theologians grappled with the intricacies of the Trinity and other points of Christian doctrine.

What I propose to do is to examine the quest for Christ by a number of figures spanning the twelfth century. I am using the term 'Christ' here in the widest sense to denote not just the figure of Jesus Christ or the second person of the Trinity. I am using the term also to cover the whole Christian understanding of God. Some of the men I have chosen wrote Christian–Jewish disputations, some were involved in crusading, some took part in the period's spiritual renewal. A number of them had connections with more than one of these activities and the lives and/or work of quite a few of them interconnect. Our purpose will be to discover whether their intellectual and spiritual quest for Christ can help us understand any better the phenomenon of central medieval persecution of Jews.

Late eleventh- and twelfth-century scholarly engagement with the problem of *Cur Deus-homo* prompted renewed interest in combating the Jewish rejection of Christ. The spate of Christian–Jewish disputations which began to appear did not just contain hackneyed biblical arguments from antiquity. Scriptural evidence was used creatively by applying to the Bible skills acquired from studying the Trivium (the three liberal arts: grammar, rhetoric and logic). Over and above that, reason was called upon to prove the truth of Christianity and the falseness of Judaism. The classical reason used by these scholars was strongly Christianized into a spiritual concept. It was seen as a route to God which helped the existing route of faith, or complemented it, or ran parallel to it. Twelfth-century thought was marked by a remarkable optimism about the possibilities of reason. Intellectuals like Anselm of Canterbury and Abelard really thought that, if used rightly, reason could yield tremendous understanding of God and his creation. But because reason was seen as a gift of God and that God was the triune God of Christianity, these thinkers in their optimism did expect reason to take humans not just to the global perception of the divine but the specific perception of Christ, son of God and God-man. When Jews were less than impressed with all the rational arguments on offer to prove the triunity of God and the divinity of Christ, scholars engaged in the Christian–Jewish debate quickly concluded that there must be something lacking in Jewish reason. Nor was this a trivial matter. The classical texts which taught them about reason also taught them it was reason that separated man from animal. A very early expression of this is in Odo of Cambrai's Christian–Jewish disputation, written between 1106 and 1113. In the final section of the disputation, where Odo argues vociferously for the rationale of the Virgin Birth, he questions his fictitious Jewish counterpart's capacity for reason. Relying on the Platonic dichotomy between spirit and appetite, he avers that only unspiritual beings like Jews, who care only for sensual things and are misled by the appearance of things, could fail to look beyond the messy insides of a woman to perceive the glory of the womb God had chosen to adopt flesh.[3] Odo, who was a realist in philosophical terms, and who had abandoned his teaching career in Tournai in order to set up an ascetic monastic house, clearly divided people into those who were spiritually inclined and those who were not. Proper use of reason opened up the spiritual world. Material things were part of the sensual world. This was the world Jews inhabited with – it should be added – profit-seeking peasants. Reason and spirit were the realms of Christians.[4]

Another scholar of the twelfth century who was determined to prove the veracity of his belief in Christ against Jews, Muslims and heretics was Peter the Venerable, abbot of Cluny (d. 1156). For Peter, Jews were even worse than Muslims because they disparaged the Virgin Birth.[5] In his lengthy diatribe against Judaism Peter repeats again and again that to his mind Jews are animals because they refuse to accept all the rational arguments Christians offer them.[6] He concludes that it is the Talmud that closes the minds of Jews to reason.[7]

Twelfth-century Christian scholars did not just formulate ideas about Jewish lack of reason on account of the fact that they did not employ their reason to recognize Christ as God. With all their interest in human motivation they began to question the traditional assumption that New Testament Jews had killed Christ in ignorance. As Jeremy Cohen has demonstrated, biblical glosses of the period begin to express the opinion that these Jews had acted deliberately.[8] They knew Jesus was God and had killed him anyway out of jealousy. They did not want to share salvation with the Gentiles. Peter Alfonsi is the first author of a Christian–Jewish disputation to make explicit the accusation of what amounts to deicide. He says in so many words that the leaders of the Jews deliberately accused Jesus of practising magic and put him to death to save their own position.[9] Peter's *Dialogue* (1108–10) was written in order to justify his own conversion to Christianity and he used his own knowledge of Jewish sources to attack Judaism in ways it had not been attacked before. It was he who introduced Christians to the aggadic (narrative) passages of the Talmud in order to demonstrate how irrational Judaism was. For Jews this work was devastating. They emerge as (1) irrational, (2) deliberate killers of God and (3) in possession of post-biblical writings which induced them to have irrational beliefs about God. These points are important because the *Dialogue* was very widely disseminated all over Europe in the twelfth and thirteenth centuries.[10] Another important contribution by Peter to the Christian–Jewish debate was his claim that a Jewish source called the *Secreta secretorum* revealed that the Tetragrammaton (the four-letter name of God) enclosed proof that God is one and three.[11] Manuscripts of his dialogue include a diagram showing how three circles containing the letters *yod*, *hey*, *vav* and *hey* overlap to create one circle. This proof was borrowed freely by many anti-Jewish polemicists, including Peter of Blois at the end of the century.[12] We shall see shortly how attractive it was to Joachim of Fiore.

Peter Abelard (d. 1142) is widely quoted for his sympathy for Jews. He speaks graphically of the difficulties of being Jewish in his *Dialogue between a Philosopher, a Jew and a Christian*, written in the late 1120s. In the same work he makes quite plain through the words of the Philosopher that he respects the zeal with which Jews try to serve God.[13] In his *Ethics*, which were written around a decade later than the *Dialogue*, Abelard suggests that, technically speaking, the Jews of the New Testament did not sin when they crucified Jesus. Consent to evil, which implies contempt of God, is what defines sin. The Jews thought they were serving God when they put Jesus to death.[14] What Abelard's contemporaries, and some modern-day scholars too, failed to notice is that there is no doubt in Abelard's mind that the Jews did something very wrong in crucifying Jesus. Having a good intention is not enough. Good intentions do have to accord with God's intentions to generate good acts. And that seems to me to be the crux of the matter. For how does anyone know what God wants him to do? To have knowledge of this a person needs to get close to God; a reciprocal loving relationship needs to develop between God and man. Two parallel paths help man to achieve this: reason and faith. Through reason man has grasp of Natural Law, that unwritten law engraved on our hearts by way of reason. Natural law has encoded in it vital lessons about loving God and man. Through faith a Christian has in addition before him the example of Jesus whose whole life, according to Abelard, was the epitome of right living and right loving of God. On top of that the Gospel of Christ comprises a comprehensive restatement of Natural Law, offering Christians all they need to formulate good intentions which are right. So what about the Jews? In his *Theologia Christiana* and in his *Dialogue* Abelard makes it quite clear that he did not think that the Law of Moses was particularly helpful in teaching its adherents how to love God rightly. Indeed all the rules and regulations of Jewish ceremonial life made Jews lose sight of what really mattered. Abelard is quite explicit in stating that real philosophers like Plato could find God more easily through Natural Law than the Jews through Mosaic Law. So for all Abelard's empathy for Jews, there can be no doubt that he was convinced that Jews did not actually get it right. Their adherence to the Law of Moses, which to his mind concerned only outward observance, ran counter to reason and the lessons of Natural Law.[15]

What we see emerging from twelfth-century anti-Jewish polemics such as these is how strongly reason was Christianized and how it could be used as a weapon against non-Christians. Classical texts taught these thinkers that all human beings shared reason, or, to put it more sharply, that human fellowship was based on the reason humans shared.[16] The Christianization of the concept of reason meant that the construct of a philosophically based universality of man could easily seem to overlap with the human fellowship on offer to all humans through the gift of Christ. Christians believed that God became man to redeem man from damnation. God-man was a brother to all men and all men were invited to share universal human fellowship through Christ. By rejecting Christ Jews could be seen not just to reject human fellowship in a religious sense but human society in a philosophical sense too. Or, to put it differently, Jews could be regarded not just as non-Christians but as less than human. This is not to say that all twelfth-century anti-Jewish polemicists consciously drew this stark conclusion. Abelard certainly did not, but Peter the Venerable did. But even Abelard questioned Jewish usage of reason and understanding of Natural Law. And it is this kind of questioning, the suggestion that Jews were somehow failing as human beings because they were Jews, that became part of central medieval anti-Jewish rhetoric, even after the concept of reason had lost much of its specifically Christian flavour. Let us now focus our attention on some thinkers who were particularly active in seeking God-man or Christ in the context of Reform.

An obvious person to start with is Rupert of Deutz (d. 1129), a man who has left us in no doubt about his views on Jews. Rupert was a black monk who was a very keen reformer in the sense that he strongly supported papal efforts to eradicate simony. He had little time for new forms of monasticism like Norbert of Xanten's order of Prémontré. But he was filled with a great need for spiritual fulfilment which he found in the figure of the crucified Christ. His copious writings pulsate with his identification with Christ; Christ's salvific blood runs off the pages of his massive biblical commentaries and his treatise against the Jews of 1126.[17] Rupert's world is the world of universal Christendom. All are invited to partake in Christ's feast. All who come with true faith in Christ will be saved. He sees monks as replacing the martyrs and confessors of the past and taking a leading role in reforming the Church. Rupert sees Jews as hateful beings who in the past persecuted Jesus out of jealousy and who in the present wickedly refuse to heed the preaching of his Gospel. Jewish hands are tainted with the blood of Christ. Jews look down on all other people. They wanted and still want redemption only for themselves; they begrudge non-Jews salvation. In fact, according to Rupert, they are not really Jews. True Jews are those who believed in their hearts the prophecy of Christ. What rolls out of Rupert's writing is not just antipathy for Jews. Jews are identified as a negative force in his vision of Christian society. The messiah Jews yearn for is none other than Antichrist. In his *Commentary on the Apocalypse* Rupert used the figure of Antichrist to embody everyone who was antagonistic to what he considered good. Jews join bad rulers, simoniacs, magicians, poets and philosophers in his Antichrist imagery.[18] In short, Rupert's dislike for Jews is an integral part of his view of reforming Christian sociey.

Connected to Rupert is the Jewish convert Hermannus, the former Judah ben David of Cologne. This is not the place to rehearse whether or not his autobiography, the *Opusculum de conversione sua* written *c.* 1150, is genuine.[19] Together with other scholars I have argued that it is real in the sense that the text reveals how Hermannus *post factum* looked back on his road from Judaism to Christianity. The connection with Rupert is that Hermannus claims to have been engaged in a disputation with him. For us the interesting points of Hermannus' work are two: (1) Hermannus categorically denies the efficacy of learned Christian–Jewish disputations in converting Jews. All the splendid arguments which Christians like Rupert produce will fall on deaf ears as long as his former co-religionists shut themselves off from the truth by a veil of blindness; (2) what converts Jews are the good deeds and prayers of converted Christians. The work is particularly fascinating in how it interlinks Christian inner conversion with the conversion Christians sought for Jews. Hermannus says that he himself was finally converted with the help of the prayers of two female recluses. He wrote his autobiography for the women and men of the Premonstratensians, the order he himself joined. Hermannus paints us a picture where attitudes towards Jews and hopes for their conversion were very much part and parcel of the ideal of a reformed Christian society.[20]

This takes us neatly to some of the figures who popularized the new spiritual goals of that reformed society. Let me start with someone who to my knowledge has not been studied in connection with Jews: the Cistercian Aelred of Rievaulx. Aelred was born in 1101 in Hexham in Yorkshire. After serving as a seneschal at the court of King David of Scotland he entered Rievaulx. He served as master of the novices from *c.* 1142. In 1143 he was elected abbot of St Lawrence of Revsby near Lincoln; he returned to Rievaulx as abbot in 1147. He died in 1167.

Aelred wrote a short book in which he used an episode in Jesus' life to explore the interior journey humans need to make through Christ in order to recover the image of God in themselves. Recovering God's image in oneself was an Augustinian idea, which became extremely popular throughout the twelfth century, especially among the Cistercians.[21] The episode is Jesus' disappearance in Jerusalem for three days (Luke 2: 42–51); the book is 'Jesus as a boy of 12', and it was completed before 1157. From the fourteenth century it circulated under the name of Bernard. The text is a devotional guide, offering its readers help in finding communion with Christ's divinity through contemplating his humanity.[22] What strikes us in the historical section of the treatise is Aelred's amazing immediacy to Jesus. Where was he during the three days that he had disappeared? What was he doing? Aelred's concern with Mary is also very intimate. He shows her fitting respect but he does wonder why she is searching so anxiously for her son. Is she worried that he might be hungry or cold or might be hurt by a child of his own age? Does she not know he is God? Aelred also gently chides her for having misplaced her son in the first place! He takes tremendous care to describe her joy at finding Jesus after three days. What had caused her so much anguish was to be deprived of Jesus' delightful presence.[23]

When Aelred turns to look at the episode in an allegorical sense he asks himself what this episode in Jesus' life means in terms of the salvific history of the Church. It is here that the fundamental issue is explored of the victory of *ecclesia* over *synagoga*. By ascending to Jerusalem Jesus shows he is abandoning the synagogue and ascending to the church of the Gentiles. He does this not to destroy the Law but to complete it by adding to the Ten Commandments the double precept of charity to reach evangelical perfection. His parents' unawareness of his staying behind in Jerusalem is an allegory for present-day Jews not knowing him as Christ. Just as Joseph handed out food to the Egyptians while his brothers went hungry in Canaan, so the language of Christ's salvation is not the language of the Jews but the language of the Gentiles, whilst the Jews remained starved of the word of God. Aelred proceeds to address Jews very directly, mocking them for imagining that they might find the Christ, i.e. the Messiah, in their own midst. He derides them for thinking this is possible after losing their Temple with its sacrifices. They had Christ in their prophecies, he came and they still did not recognize him. '*Quem queritis, o Judaei? Quem queritis?*' 'Whom are you seeking, o Jews, whom are you seeking? You are dispersed over the whole world, and everywhere you offend Christ. Everywhere in praise of Christ the peoples say your amen, it is your Hosanna that resonates in praise of Christ. And still you search ... ?' Jews have rejected the gifts of Christ. Jealousy prevented them from recognizing their own saviour. But just as Jesus was found after three days the Jews will in the end of time convert after the suitable ages of the Church have passed.[24] This sentiment ties in with Aelred's fifteenth sermon on the burdens of Isaiah where he waxes lyrically on the existence of universal love after the final conversion of the Jews.[25]

The third part of the text takes us to the moral meaning of the story: the inward turning to Christ. This is the moral interpretation of the text about which Aelred cared most. Jerusalem here becomes the image of contemplation. The crescendo comes when '[Christ] will come to you in the aroma of perfumes and spices, he will imprint on your soul a celestial and divine kiss and fill your entrails with ineffable sweetness and you will cry in delight "grace is poured abroad in thy lips(Ps 44: 3)".' Aelred uses the image of the female sinner who kisses and anoints Jesus' feet (cf. Luke 7: 36ff.) to show how he is the refuge for the whole world. He despises no one; no one is rejected. It is there that the leopard changes his spots. The only one who does not change is the Pharisee in whose house the scene takes place. As far as the Pharisee is concerned, the unction reeks of death. But it is the Pharisee, who is worried about coming into contact with her sin, who stinks through the tumour of his pride. Pharisaic (i.e. Jewish) pride and stand-offishness is contrasted with Christ's ubiquitous mercy.[26]

So what do we have here? We have a work which amounts to a passionate love story about Christ written in beautiful rhythmic Latin, which culminates in the Christian believer experiencing Christ as part of himself. And part of this story, which really has nothing whatsoever to do with Christian–Jewish relations, uses acrimonious language to condemn Jews. Condemnation of Jews functions here as an integral part of the unfolding of Christian spirituality. Condemnation of contemporary Jews goes hand in hand with the expectation for the final conversion of Jews when all peoples, Jews and Gentiles, will serve each other in charity within the lap of the Church.[27]

Let us proceed with another Cistercian who predated Aelred by a few decades: William of St Thierry. William is another theologian who is not usually studied by those interested in the Christian–Jewish debate.[28] He was born in the last quarter of the eleventh century in Liège. He studied at Liège and Rheims but chose to enter monastic life in St Nicaise of Rheims, which had recently been reformed by the Benedictines of La Chaise Dieu in Auvergne.[29] In *c.* 1119 he became abbot of St Thierry near Rheims. He must have met St Bernard towards the end of 1118. Years of inner conflict followed in which he felt drawn to the austere contemplative life of the Cistercians. At the same time he saw it as his abbatial duty to help to bring about renewal in the Benedictine order. Eventually he left his monastery to become a Cistercian at Signy in 1135.[30] He stayed there until his death in 1147 or 1148. He had written a number of works before entering Signy, for example, *On Contemplating God*, but much of his *oeuvre* stems from his Signy days, including his *Exposition of Romans*, his *Commentaries on the Song of Songs*, *Mirror of Faith*, *Enigma of Faith* and his best-seller: *The Golden Epistle*, his letter to the Carthusians at Mont Dieu. This work was referred to over and over again by those concerned with spirituality. Its popularity probably owes much to the fact that from the second half of the twelfth century it circulated under the name of St Bernard.[31]

In this work William offers a kind of DIY kit for loving God. As with Aelred, we are entering here the inner contemplative world of the Cistercians. As with Aelred, man is searching here to regain the image of God he lost in the Fall. Ultimate love of God denotes knowledge of God. When this ultimate love has been painstakingly achieved, man will become united to God. He will have progressed from *amor* (a strong inclination to God) to *dilectio* (a clinging to God) to *caritas* (enjoyment of God) to *unitas spiritus* (willing only what God wills).[32] But let us start at the beginning. In his letter William explains how man first has to rise above his animal self. Obedience is the key here. The new monk must observe perfect obedience. Perfection is achieved when habitual virtue has become a pleasure. At this stage it is vital for the monk to recall and contemplate Christ's Passion. William recommends that he spend at

least one hour a day attentively concentrating on the benefit which was
bestowed on man by the Passion. He describes this thought process as
spiritually consuming the body and blood of Christ.[33] The next stage is
the rational stage when man begins to understand the teachings of faith
he is obediently following. This stage reaches perfection and passes into
the third and final stage when the judgement of reason passes into spiritual affection. Progress in that phase is 'to look upon God's glory with
face uncovered; its perfection is to be transformed into the same likeness, borrowing glory from that glory, enabled by the Spirit of God'.[34]
'In a manner which exceeds description and thought, the man of God is
found worthy to become not God but what God is, that is to say man
becomes through grace what God is by nature.'[35] And towards the end
of the letter he says: 'So eventually love arrives at some likeness of that
love which made God like to man by accepting the humiliation of our
human lot in order that man might be made like to God by receiving
the glorification of communion in the divine life.'[36] The possibility of
man participating in God in William's thought is – to say the least –
remarkable. William does not speak of Jews in this text but it is worth
our while to ask what this kind of specialized spirituality, which is
grounded in Christ and the Holy Spirit, might think of non-Christians
and their ability to love God. For this we have to examine some of
William's other works, which were not as widely disseminated as his
letter to the Carthusians.

As Hourlier has noted, William is explicitly pessimistic about the
spiritual chances of non-Christians. Philosophers do not know God,
they do not have faith, they do not love God, they cannot find beati-
tude, they are incapable of the least bit of justice, they can only err.[37] In
the treatise *On Contemplating God* he says plainly that those who did not
love God, did not love righteousness. They may have had a kind of love
and they may have performed some honest actions but they did not
have the faith that works by love.[38] 'Love' in William's thought is a
'vehement will' or a 'vehement well-ordered will'. It is a force that takes
the soul to its appointed goal and that goal is God.[39] Thus in his *Exposition on the Song of Songs* William is adamant that no virtue is a true
virtue unless it allows its possessor to 'savour him who is the Wisdom of
all virtues', i.e. Christ. 'As true and solid virtue is found in him (not
merely zeal for labor or ambition of will, but the soul's inclination and
the habit of a well-disposed mind), so when love is directed toward
him, it is true and living.'[40] In his *Mirror of Faith* William claims that
'unbelievers love to err. They have loves which they do not want to
master, and through them they are sent into error from which they do

not know how to return. The sense of the flesh is in the persons who do not know how to think about spiritual things except in a physical way, much less do they know or can they think about those things which concern the human dispensation in the Lord Jesus Christ except carnally.'[41]

William's *Exposition of the Song of Songs* tells the adventure of the soul that has conquered sin and clinging to Christ becomes the spouse. The soul achieves spiritual union with God through Christ and with the help of the Holy Spirit. The Holy Spirit is the reciprocal love between Father and Son in the Trinity and their mutual knowledge. But it is also the love instilled in man by God. To use Bell's words, 'God's love for us and . . . our love for God . . . are [both] the Holy Spirit, who is himself the consubstantial love of the Father and Son.'[42] By inhabiting man, the Holy Spirit unifies man to God. Borrowing from Origen, William traces this spiritual adventure from the enticement of love of the simple Christian to the final phase which is the spiritual state in which the soul has become the spouse. The simple Christian is in the first phase; he is what William calls the animal man, the man who

> prays to God, but he knows not how to pray as he should. . . . It does not satisfy him to think of God or look upward to him as he is; he cares only that he has the power to grant what is asked for . . . this man, multiplying words in his prayer sometimes without comprehension, thoughts without understanding, does not seek comprehension of God or affection for him. Even if these should present themselves to him as it were spontaneously, he turns them into something else. For him, therefore, as of old for Israel according to the flesh, the God to whom he prays is ever in the dark cloud.[43]

It is clear that in William's thought man cannot even start looking for God without faith in the Trinity. In his *Enigma of Faith* William is quite frank about the fact that the dogma of the Trinity is hard to understand. He writes that from the time of Jesus belief in one God was not a problem. According to him all were in agreement about that, Jew and Christian, pagan and barbarian. It is the Trinity that is what worries people. Is this god not three gods? This question constitutes the enigma of faith. It scares the impious and makes them flee from the face of the Lord. But it is beneficial to the pious for it stirs them up and urges them on to seek his face always. In other words, it is the Trinity that sets man going on his spiritual adventure. It prickles him into spiritual activity.[44]

This is not the place to delve any further into William's highly complex theological thinking. Nor would it serve our purpose to enter the existing discussion among scholars about what he borrowed from Augustine or the Greek Fathers and what he owes to St Bernard and vice versa.[45] What strikes us are two points. The first one is how exclusively Christian the content of his spirituality is. William's engagement with Christ and the Holy Spirit forces us to realize how essential an exclusively trinitarian understanding of God was for this kind of spiritual thought. The second point is that we see in William's thought how the image of carnal Israel can become an integral element of Christian spirituality. Carnal Israel no longer only denotes a stage in the salvific history of the church, with carnal Israel prefiguring the coming of Christ and his redemption. Carnal Israel has become a Christian category in a Christian's ascent to God. Why is this important? With carnal Israel prefiguring the true Israel of Christ, Jews could be said to play a beneficial role in Christian society in that they bore witness to the truth of Christian prophecy. This is of course the substance of Augustine's concept of *Testimonium Veritatis* (witness to the truth).[46] But with carnal Israel signifying an internalized Christian characteristic it becomes much harder to define clearly what role flesh-and-blood Jews need to play in Christian society. There are enough unspiritual Christians about to emphasize the need for spiritual development and reform. This kind of thinking not only denigrates Jews by emphasizing over and over again their lack of spirituality, it not only uses Jews to express what Christians should leave behind them, it begs the question whether Christian society needs their service at all.

This line of enquiry can, I think, also shed some light on the much studied St Bernard.[47] In his very well known work *De Diligendo Dei*, which was written around 1126, Bernard teaches how man should learn to love God. In the section of the work where he explores the causes for loving God he says that even infidels have no excuse for not loving God with all their heart, soul and might. For they are able to recognize the deity as the author of human dignity, bestower of knowledge and giver of virtue. In other words, the admonition of Deuteronomy 6: 4 to love God made sense before the coming of Christ. But within seconds we are told that it is not possible for man to know how to love God properly by his own powers. Man's sinful pride will make him seek his own glory rather than God's. Christians are tremendously helped in this area for they know how much they need Christ. They are pricked, as Jews and pagans are not, by Christ's Passion.[48] When he comes to explain what the different degrees of love are which lead to the *summum* of love for God, he discusses the first rung as the love of animal or carnal man.

This animal love is implanted by nature. When it inclines towards God it does so for the sake of man, not God. As we have seen above, animal man loves God because he realizes that he gets benefits from God. This is selfish carnal love, which can and should grow into love of neighbours and through that more altruistic love to love for God for God's own sake. The role of Christ's love for man, the inducement to love as Christ loved, is absolutely vital in the process.[49] As we saw with William the inner logic of Bernard's scheme has him connect the first step in a Christian's development with carnal animal man. Although he does not specifically link the term carnal to the term Israel, as William did, it would seem to me that there can be little doubt the two men are referring to the same thig. If I am right, we can then try to understand some of Bernard's anti-Jewish pronouncements as not just reflecting his straightforward love for Christ and his sincere belief that the church superseded the synagogue but as something connected to his spiritual programme.[50]

In Bernard we have the rare opportunity to put together Christian theory and practice about Jews side by side. In two instances he had to deal with real Jews or, in the case of one of them, at least with someone of Jewish descent. As is well known, Bernard stopped the Cistercian Ralph from preaching violence against Jews at the time of the Second Crusade. David Berger has pointed out that Bernard's action here was not without risk. Ralph was popular in Germany and Bernard could not be sure he could stop him.[51] Jeremy Cohen has argued convincingly that Bernard's rejection of Ralph's violent message against Jews was part and parcel of his own Christian *Weltanschauung*. It had, in fact, very little to do with love for Jews.[52] It is clear that Bernard would not countenance such a flagrant transgression of the core principle of *Testimonium Veritatis*. However strongly he condemned Jewish rejection of Jesus Christ, he still adhered to the basic Christian tenet that in the end of time Jews would convert. In the meantime they must not be attacked. Jews simply did not fall into the same category as Muslims, who had taken up arms against Christians. Added to that, Ralph's independent activity hardly matched Bernard's demands of his reformed monks. But there is more to it than that. Even as Bernard argued against attacking Jews, he connected them to one of the sins reformers like himself were so opposed to: usury. He did not only assert that crusaders should not be charged interest by Jewish lenders. He internalized the concept of being Jewish in a Christian sense by making the word *judaisare* mean 'lending money on interest'. Christian moneylenders were to his mind tantamount to baptized Jews.[53]

The adoption of the notion of Jewish behaviour here as an internal pejorative Christian concept might help us understand better the way Bernard and others reacted to the contest between Popes Anacletus II and Innocent II in the 1130s. Anacletus was Petrus Pierleoni and of Jewish descent. Innocent was the Papal Chancellor Haimeric's candidate and supported by the rival family to the Pierleoni, the Frangipani. Haimeric is the man to whom Bernard dedicated his *De Diligendo Dei*. At the beginning of the controversy Anacletus' position was strong. He had support from the majority of cardinals. Haimeric was therefore forced to fight the battle on the grounds that Innocent was the better candidate of the two. He put him forward as the obvious choice of reformers and Anacletus as the antithesis to what they espoused. His campaign made much of the fact that (1) the Pierleoni stemmed from a converted Jew, and (2) the Pierleoni were papal bankers and heavily engaged in usury.[54] Much ink has been spilt over the Anacletus affair and the anti-Jewish invectives it generated.[55] What concerns us here is the way usurious activity and 'being Jewish' seem to be intertwined in people's minds. It seems to me that when Bernard wrote that it was 'to the injury of Christ that a man of Jewish race has seized for himself the see of St Peter' we are not just faced with straightforward antipathy for Jews being used against the grandson of a Jewish convert.[56] I think Bernard was also using the concept of being Jewish in an internalized way to lampoon the well-known usurious activities of the Pierleoni family. The same could perhaps be said for Orderic Vitalis, who wrote so scathingly of Anacletus' brother at the Council of Rheims in 1119. Gratianus is described by Orderic as being physically deformed, looking more like a Saracen or a Jew and being scorned on account of the hatred felt for his father [who was bon a Christian] who everyone knew was an infamous usurer.[57] Other opponents of Anacletus, like Arnulf of Seez, accused him of looking Jewish, having incestuous relations with his sister, and coming from a family not purified of the yeast of Jewish corruption.[58] It seems to me that much of the anti-Jewishness being expressed by Anacletus' opponents relied heavily on the internalization of the concept of being Jewish. What they saw or wanted to see in Anacletus was not only bad Christian behaviour which they identified with Jewishness. They saw that combined with real Jewish ancestry. The Christian internalization of Jewishness would obviously make it all the harder to accept that real Jews could fully abandon Jewishness in the sense of primitive, unspiritual or bad behaviour when they converted or, as Arnulph put it, expiate the ferment of Jewish corruption.[59] When converts or their descendants were engaged in the activity which was considered to be

Jewish *par excellence* it could seem to prove that they had remained Jews.

The internalization of carnal Israel is also a marked feature of the crusading propaganda of someone like Peter of Blois (d. 1211/12), the famous letter writer who worked for the Cistercian Archbishop of Canterbury, Baldwin, and who preached the Third Crusade. Going on crusade is preached by Peter very much as an act of inner conversion. The fall of Jerusalem, the loss of the true cross, all of this means that God is punishing Christendom for the enormity of its sins. Taking the cross is the way of true penance. God will turn his face back in favour on Christians if they reform themselves and turn to Christ. Peter's crusading works belabour the themes we have touched on in our discussion of Christian reform: inner conversion, zeal for Christ, and love for God. And throughout his works he uses the concept of carnal Israel in a Christian sense. Carnal Israel does not denote Jews. Carnal Israel denotes misguided Christians. The language he uses to arouse the Christians of his day to crusade is the prophetic language used in the Bible to arouse the Israelites of old. Again the question presses: if this is how carnal Israel is perceived, what role is there left for real flesh-and-blood Jews? In Peter's work we can actually put our finger on a close connection between these ideas and hostility to contemporary Jews. His anti-Jewish polemic (written *c.* 1196) is not only marked for his harshness against Jews, it is the first to contend that Jews commit ritual murder.[60]

Zeal for reform and/or crusade did not necessarily predispose thinkers to be positive about anti-Jewish polemics. We have already seen how little regard Hermannus had for anti-Jewish disputations. Another example of this attitude is the reaction of Adam of Perseigne to an invitation to 'write something with which the faith of believers could be confirmed and the obstinacy of unbelievers, especially the Jews, could be confuted'.[61] Adam was born *c.* 1145; he probably spent some time at the court of Champagne before becoming a regular canon. In the course of his conversion he proceeded first to join the Benedictine order and then to become a Cistercian. He became abbot of Perseigne in *c.* 1188 and died in *c.* 1221. Adam is well known for his letters and sermons and for the missions he undertook for the papacy. In 1195 he was asked to look into the ideas of Joachim of Fiore. He preached the Fourth Crusade from 1201 and went on crusade himself. In 1208 he was charged by Innocent III to broker peace between Philip Augustus and John so that they could support the Albigensian Crusade. Adam was an ardent reformer and a great admirer of Peter the Chanter. He supported the preaching movement inspired by Peter and Fulk of Neuilly at the end of the century in

Paris. But when the preaching against the evils of usury extended itself to anti-Jewish activity and a court priest (court of Champagne?) involved in Fulk's campaign against the Jews asked Adam *c.* 1198 to supply him with suitable material, he turned down the request in exceedingly harsh terms.[62] Adam berates his correspondent for wanting to combat Jews while he himself is so utterly devoid of the qualities his sacerdotal condition demands of him. According to Adam, disputing Jews was an obvious waste of time. God in his justice had punished Jews with internal blindness. Their hearts are obdurate and until the day that the plenitude of peoples would enter the kingdom, they will remain blind. Discussing truth with Jews constitutes a blindness equal to theirs. People who take on debates with Jews do so for their own vainglory. Priests should know better than to indulge in such inane vanity. They should seek to regain the image of God's likeness. They should not just mouth Christian doctrine, they should practise what they preach and be an example of good Christian living. Their unrighteous behaviour and impious handling of the sacrament make them worse than Jews. The Jews killed Jesus out of ignorance; these Christians were killing the Christ they feigned to worship. Adam ends by claiming that his correspondent should leave disputing with Jews alone, that is not his business. Reforming Christians and himself, who are worse than Jews, through word and deed, is.[63]

What emerges from Adam's letter is a passionate concern that the clergy should not be tempted away from their pastoral duties to any other activities, however exciting they might seem. We can see another expression of this attitude in the letter he wrote *c.* 1214 to Hamelin, bishop of Le Mans, about preaching a new crusade. In it he argues strongly that clerics should stay at home. That is where their work lies. 'For Christ paid the price of his own blood not for the acquisition of the land of Jerusalem but rather for the acquisition and salvation of souls'. Adam is clearly worried about the motives crusading clerics might have in deserting their pastoral duties for the sake of travelling to the east. In his reply to Pope Innocent III's letter canvassing bishops and abbots about their ideas concerning crusade and reform in the planning stages of Lateran IV Adam avoids any mention of crusading. His reply touches only on the reform of the regular and secular clergy. Andrea has connected Adam's scathing criticism of crusading clerics to his disillusionment with the Fourth Crusade.[64] Taking his other letters into account, including his 1198 letter about the Jews, one could say that Adam was clearly a reformer whose individual spiritual quest for Christ inspired him to admonish others to serve Christ in the way that he sought to himself. He seemed especially concerned about the actions of priests who were charged with the spiritual care of others. Priests should preach by example and not just through words. This is the context of his concern about the proper sphere of action for clerical reformers. Jews were clearly not part of his agenda; Christians, especially priests, were.[65]

A one-time Cistercian who did think the time was right for Jewish conversion and who did write an anti-Jewish polemic was Joachim of Fiore. Joachim was born *c.* 1135 in Celico in Calabria. He travelled to the Holy Land as a pilgrim *c.* 1167 while he was a junior chancery official. The experience confirmed his desire to live an ascetic life of preaching. He became a monk in Corazzo around 1171. As prior and abbot of Corazzo he worked hard to achieve the monastery's incorporation into the Cistercian order. In 1183 Joachim was at Casamari in an attempt to convince the abbey to take on Corazzo as a daughter house. It is at Casamari that Joachim experienced revelations of the Trinity. Years of writing followed which left little time or interest on Joachim's part for Corazzo. In 1188 Pope Clement III arranged for Corazzo to achieve Cistercian status and excused Joachim from his position as abbot so that he could continue his writing. Soon after that Joachim founded a monastery in Fiore to create the Florensian order, which would follow a rule that was even stricter than the Cistercian interpretation of it. He died in 1202.[66]

Joachim imagined the course of history to follow the pattern of the Trinity. He taught that the world would pass through three statuses, that of the Father, the Son and the Spirit. According to him the third status of the world was at hand. Unlike other eschatological thinkers, Joachim placed the era of the Spirit on earth before the end of time. In a manner that is sometimes reminiscent of Rupert of Deutz's thought, he expected this to be a period in which reformed monks would take on the role of spiritual men leading a spiritualized reformed universal church. Conversion of the pagans would herald conversion of Jews at the start of the era. Out of the complete concordance between the Old and New Testaments a new spiritual sense of scripture would become the hallmark of the third spiritual age.[67] In his tract against the Jews Joachim says quite plainly that he is sure the time of mercy for the Jews is at hand.[68] The polemic is a remarkable piece. It lacks the harsh vituperative language characteristic of other twelfth-century Christian–Jewish polemics. It is plain that Joachim knew Jews in his native Calabria. It is likely that his tone reflects the relatively good relations between Christians and Jews in this period in southern Italy.[69] It is also possible that he shaped it so that it really could be an effective tool in the hands of Christians engaged in converting Jews.[70] His hope is that some Jews will be saved even before the imminent general conversion of the Jews.[71] In it he seriously addresses Jewish criticism of Christian beliefs and genuinely tries to find common ground between Judaism and Christianity. For example, when discussing the Trinity he asserts that both religions share common belief in God the Father; the Son and the Holy Spirit are the areas of contention.[72] A bit later he really tries to come to terms with Jewish distaste for the concept of God taking on flesh: 'when we Christians say God had a son, you must not imagine anything human or corporal in your hearts...'.[73] He does not belabour the idea of Jewish responsibility for the death of Jesus Christ.[74] The whole work is aimed to show Jews that a spiritual reading of the Hebrew Bible would prove the truth of Christianity. His purpose is crystal clear. It is to convert Jews. Jewish conversion is an absolutely integral element of his prophecies and in that sense he apportions a positive role to Jews.[75] But the positive role he assigns to Jews is exclusively connected to their conversion, which will serve the destiny of mankind. Jews outside the context of conversion were part of the enemy camp consisting of those who followed the letter rather than the spirit. Spiritual understanding of both the Old and the New Testaments was vital for Joachim. The age of the spirit would bring about a full understanding of Scripture. Joachim used harsh words for those like the Jews who followed the letter.[76] As he says in his *Exposition on the Apocalypse*: 'The external evil [the Jews] did [when they crucified Jesus] was a sign of the greater evil they conceived within, that is, to snuff out the spiritual understanding and bury it in the belly of the letter so that its voice might be heard no more in their streets nor have any further place in their possessions.'[77] But in the same work he expresses confidence in Jewish conversion. A section of the work is devoted to analysing the Tetragrammaton, as Peter Alfonsi had done at the start of the century. This is information Joachim did not have to hand when he wrote his anti-Jewish polemic. Joachim was spellbound by this revelation and used it extensively. He seemed to think that this could have an immediate impact on individual Jews. He claims to have tried it on a Jewish friend, who apparently was willing enough to accept that Alfonsi had expressed these musings, but not to convert.[78] This setback did not, however, seem to dampen Joachim's enthusiasm for, and confidence in, the imminent general conversion of the Jews to Christ. That that would occur seems to have been a given for Joachim.

Not everyone was enchanted by the fanciful eschatological musings of Joachim of Fiore. He was censured by the Cistercians in 1192 as a fugitive from their order. In the 1190s Abbot Geoffrey of Auxerre accused him of Jewish descent. At the Fourth Lateran Council he was condemned for his quarrel with Peter Lombard's work on the Trinity. The papal commissioners at Anagni in 1255 were critical of aspects of Joachim's work, including his predictions about Jewish conversion.[79] But later Cistercians and especially Spiritual Franciscans were attracted to the role he assigned spiritual men in the third status. And many of his ideas attracted a wide following.[80] For our purposes the most important point is Joachim's firm belief that Jews were about to convert. What is probably the most positive anti-Jewish polemic that we have of the twelfth century hinges on the author's steadfast conviction that Jewish conversion really is at hand.

The political, socio-economic and ecclesiastical developments of the central Middle Ages went hand in hand with a growing sense of Christian self-importance. Princes and their advisers thought hard about the specifically Christian nature of their Christian body politics. Theologians and lawyers deliberated about the Christian morality of a society undergoing the effects of an evolving money economy. Churchmen and scholars adapted traditional modes of thought and ecclesiastical structures as they discovered new texts and explored new forms of spirituality. A particular focus of this period was the figure of Jesus Christ. Identification with Christ did much to shape the ideas of Christians about themselves as human beings and their role in society. And an integral part of their Christian self-consciousness was the need to rethink their relationship with their fellow Jews. Augustine's legacy of *Testimonium Veritatis* did, of course, present Christians with a traditional model to follow in this respect, but many of the new ideas of the period did seem to question the very nature of that concept. Whereas the model allowed the Jews to play the part of servants in Christian society, new ideas about the essence of the fellowship of man and the nature of Jews questioned whether Jews were really part of the human family. Other notions posited Jews as actual enemies of humanity because they were scripted as deliberate killers of Christ and because their post-biblical literature supposedly corrupted their reason and faith. Still others used Jews and Jewishness to epitomize what they thought was wrong about Christian society. Others took this kind of thinking a step further by internalizing the meaning of carnal Israel and Jewishness to denote spiritually void or undeveloped Christianity. Still others were so imbued with the need for Christian conversion that they sought for immediate Jewish acceptance of Christian spiritual truths. One can almost say that support for the principle of *Testimonium Veritatis* came from those who had real doubts about the motivations underlying persecution of Jews and from those who were concerned about the social and economic changes which persecution of Jews would cause. Nor, in the final analysis, were most spokesmen of the institutional church willing to forgo the expectation of the apostolic promise that at the end Jews would, indeed, acknowledge Christ.

In most cases it would be hard to argue for a direct link between the ideas of any one person we have studied and a particular instance of persecution against Jews. On the other hand, it seems clear that by delving into the very heart of internal Christian thought and spirituality we have uncovered a rich reservoir of anti-Jewish thought that strongly implied that Jews, who had no intention of converting, had nothing positive to contribute to reforming Christian society. A deeper understanding of those ideas can indeed help us to understand better why this period in European history, which was characterized by intense spiritual reform and remarkable intellectual renewal, was also marked by increasing persecution of Jews.

Notes

1. J. Riley-Smith, 'The First Crusade and the Persecution of the Jews', in W. J. Sheils (ed.), *Persecution and Toleration*, Studies in Church History, 21 (Oxford, 1984) 51–72.
2. B. McGinn, *The Calabrian Abbot: Joachim of Fiore in the History of Western Thought* (New York and London, 1985) p. 7.
3. PL 160, coll. 1110–12; trans., I. M. Resnick, *On Original Sin and a Disputation with the Jew, Leo, Concerning the Advent of Christ, the Son of God: Two Theological Treatises, Odo of Tournai* (Philadelphia, PA, 1994) pp. 95–7.
4. On Odo see my *Christians and Jews in the Twelfth-Century Renaissance* (London, 1995) pp. 108–10. Realists believed in the objective existence of general ideas.
5. G. Constable (ed.), *The Letters of Peter the Venerable* (Cambridge, MA, 1967) pp. 327–30.
6. *Petri Venerabilis Adversus Iudeorum inveteratam duritiem*, ed. Y. Friedman, CCCM 58 (Turnhout, 1985) pp. 57–8, 125.
7. He discusses the Talmud in book V of his treatise against the Jews, ed. Friedman, pp. 125–87. On Peter the Venerable see my *Christians and Jews*, pp. 115–17.
8. J. Cohen, 'The Jews as Killers of Christ in the Latin Tradition, from Augustine to the Friars', *Traditio,* 39 (1983) 1–27.
9. PL 157, coll. 573, 648–9; see also J. Tolan, *Petrus Alfonsi and his Medieval Readers* (Gainesville, Fl, 1993) pp. 19–22.

10. For the manuscript tradition of the *Dialogue* see Tolan, *Petrus Alfonsi*, pp. 98–107, 182–98.

11. PL 157, col. 611. The source has not been identified.

12. PL 207, col. 833. For more on Peter of Blois, see below.

13. *Dialogus inter Philosophum, Judaeum et Christianum*, ed. R. Thomas (Stuttgart-Bad Cannstatt, 1970) pp. 49–53; trans. P. J. Payer, Medieval Sources in Translation, 20 (Toronto, 1979) pp. 30–5. See for my arguments against identifying Abelard's philosopher with a Muslim, '*Intentio Recta an erronea*? Peter Abelard's views on Judaism and the Jews', in B. S. Albert et al. (eds), *Medieval Studies in Honour of Avrom Saltman*, Bar Ilan Studies in History, 4 (Ramat-Gan, Israel, 1995) pp. 13–30, repr. as item XIII in A. Sapir Abulafia, *Christians and Jews in Dispute: Disputational Literature and the Rise of Anti-Judaism in the West (c. 1000–1150)*, Variorum Collected Studies Series (Aldershot, 1998). On Abelard see also J. Marenbon, *The Philosophy of Peter Abelard* (Cambridge, 1997) and M. T. Clanchy, *Abelard: A Medieval Life* (Oxford, 1997) pp. 245–6, where Clanchy remarks on Abelard's lack of interest in Islam.

14. D. E. Luscombe (ed. and trans.), *Peter Abelard's Ethics*, Oxford Medieval Texts (Oxford, 1971) pp. 51–67. I am following Marenbon's dating (pp. 66–7) which takes previous scholars' dating into account, especially that of Constant Mews, 'Dating the Works of Peter Abelard', *Archives d'Histoire Doctrinale et Littéraire du Moyen Age*, **52** (1985) 104–26.

15. *Dialogus.*, ed. Thomas, pp. 52–84, trans. Payer, pp. 35–71, *Theologia Christiana*, II, 44, ed. E. Buytaert, *Petri Abaelardi opera theologica*, II. CCCM, 12 (Turnhout, 1969) p. 149. On Abelard and the Jews see my '*Intentio Recta an erronea?*'

16. A cursory glance at, for example, Cicero's *De Officiis* (*On Duties*) would be enough to get this message. We know from the manuscript tradition that the text was widely available.

17. *Annulus sive Dialogus inter Christianum et Iudaeum*, ed. R. Haacke, in M. L. Arduini, *Ruperto di Deutz e la controversia tra Christiani ed Ebrei nel seculo XII* (Rome, 1979) pp. 183–242.

18. PL 169, coll. 1124–25. On Rupert see my 'The Ideology of Reform and Changing Ideas concerning Jews in the Works of Rupert of Deutz and Hermannus Iudeus', *Jewish History*, 7 (1993) 43–63, repr. as item XV in my *Christians and Jews in Dispute*; D. E. Timmer, 'Biblical Exegesis and the Jewish–Christian Controversy in the Early Twelfth Century', *Church History*, **58** (1989) 309–21; J. H. van Engen, *Rupert of Deutz* (Berkeley and Los Angeles, CA, 1983); B. McGinn, *Antichrist: Two Thousand Years of the Human Fascination with Evil* (San Francisco, CA, 1994) p. 122.

19. Ed. G. Niemeyer, MGH, Die deutschen Geschichtsquellen des Mittelalters 500–1500, 4 (Weimar, 1963).

20. On Hermannus see my 'The Ideology of Reform'. Avrom Saltman claimed the work a fake in his 'Hermann's *Opusculum de conversione sua*: Truth or Fiction?, *Revue des Etudes Juives*, **147** (1988) 31–56.

21. D. N. Bell, *The Image and Likeness: The Augustinian Spirituality of William of St Thierry* (Kalamazoo, MI, 1984) pp. 108–10; G. Constable, *The Reformation of the Twelfth Century* (Cambridge, 1996) p. 279; G. Constable, *Three Studies in Medieval Religious and Social Thought* (Cambridge, 1995) pp. 183–91; A. Squire, *Aelred of Rievaulx: A Study* (London, 1981) pp. 68–9.

22. Aelred de Rievaulx, *Quand Jésu eut douze ans*, ed. A. Hoste, trans. J. Dubois, Sources Chrétiennes, 60 (Paris, 1958) pp. 10–33.
23. *Quand Jésu*, pp. 48–51, 64–5.
24. *Quand Jésu*, pp. 76–91.
25. *Sermones de oneribus in capp. XIII et seq. Isaiae prophetae*, XV, PL 195, coll 419–22; Squire, *Aelred of Rievaulx*, pp. 141–3.
26. *Quand Jésu*, pp. 90–113.
27. PL 195, coll. 419–20; Squire, *Aelred of Rievaulx*, pp. 141–2.
28. Friedrich Lotter takes a very different view of William from mine in 'The Position of the Jews in Early Cistercian Exegesis and Preaching', in J. Cohen (ed.), *From Witness to Witchcraft: Jews and Judaism in Medieval Christian Thought*, Wolfenbütteler Mittelalter-Studien, 11 (Wiesbaden, 1996) pp. 163–85. I do not dispute that William expresses a very clear idea of the place of the Jews in Christian salfivic history in his commentaries on Romans and that he identifies with Paul's teaching concerning the final conversion of the Jews, but I cannot read the commentaries as being as positive towards the Jews, as Lotter does. A translation of William's commentaries can be found in his *Exposition on the Epistle to the Romans*, trans. J. B. Hasbrouck, ed. J. D. Anderson, Cistercian Fathers series, 27 (Kalamazoo, MI, 1980).
29. L. Melis, 'William of Saint Thierry, his Birth, his Formation and his First Monastic Experiences', in *William, Abbot of St. Thierry: A Colloquium at the Abbey of St. Thierry*, trans. from the French by J. Carfantan (Kalamazoo, MI, 1987), pp. 9–33. Scholars disagree about what he studied. Melis concludes that he either studied liberal arts at Liège and theology at Rheims or liberal arts at both places.
30. S. Ceglar, 'William of Saint Thierry and his Leading Role at the First Chapters of the Benedictine Abbots (Rheims 1131, Soissons, 1132)', in *William, Abbot of St Thierry*, p. 34.
31. William of St Thierry, *The Golden Epistle: A Letter to the Brethren at Mont Dieu*, trans. T. Berkeley, introd. J. M. Déchanet, Cistercian Fathers Series, 12 (Spencer, MA, 1971) p. xi; idem, *Exposition on the Song of Songs*, trans. C. Hart, introd., J. M. Déchanet, Cistercian Fathers Series, 6 (Shannon, 1970) pp. vii–xii.
32. *Golden Epistle*, p. xxv.
33. *Golden Epistle*, 45, 115, ed. J. Déchanet, *Guillaume de Saint-Thierry, Lettre aux frères du Mont-Dieu (Lettre d'or)*, Sources Chrétiennes, 223 (Paris, 1975) pp. 178–81, 234–5; trans., Berkeley, pp. 27, 49–50.
34. *Golden Epistle*, 45, ed. pp. 180–81; quotation taken from trans., Berkeley, p. 27.
35. *Golden Epistle*, 263, ed. pp. 354–5; quotation taken from trans., Berkeley, p. 96.
36. *Golden Epistle*, 272, ed. pp. 362–3; quotation taken from trans., Berkeley, p. 98.
37. J. Hourlier, 'St. Bernard et Guillaume de Saint-Thierry dans le "liber de amore" ', *Analecta Sacri Ordinis Cisterciencis*, 9 (1953) 225.
38. William of St Thierry, *On Contemplating God*, 12, in *Guillaume de Saint-Thierry, La Contemplation de Dieu, L'Oraison de Dom Guillaume*, ed. J. Hourlier, Sources Chrétiennes, 61 (Paris, 1959) pp. 108–11; idem, *Prayers, Meditations*, trans. Sister Penelope, Cistercian Fathers Series, 3 (Shannon, 1971) pp. 59– 60.
39. I am following here the very clear explanation of Bell, *Image and Likeness*, pp. 127–8.
40. William of St Thierry, *Exposition on the Song of Songs*, song 1, stanza 8, 105, in *Guillaume de Saint-Thierry, Exposé sur le Cantique des Cantiques*, ed. J.-M.

Déchanet, trans. M. Dumontier, Sources Chrétiennes, 82 (Paris, 1962) pp. 236–7; quotation taken from trans., Hart, p. 84.

41. *Mirror of Faith*, 60, in *Guillaume de Saint-Thierry, Deux traités sur la foi: le Mirroir de la foi, l'Enigme de la foi*, ed. and trans. M.-M. Davy (Paris, 1959) pp. 74–5; quotation from trans. T. X. Davis, introd. E. R. Elder, Cistercian Fathers Series, 15 (Kalamazoo, MI, 1979) p. 68.

42. Bell, *Image and Likeness*, p. 135, see also p. 134.

43. William, *Exposition on the Song of Songs*, Preface, 14, ed. Déchanet, 86–9; quotation from trans., Hart, p. 12, see introduction to Hart's translation by Déchanet, pp. xiii–xxxi.

44. *William, Enigma of Faith*, 67, in *Guillaume de Saint-Thierry, Deux traités sur la foi: le Mirroir de la foi, l'Enigme de la foi*, ed. and trans. M.-M. Davy (Paris, 1959) pp. 150–1; trans. J. D. Anderson, Cistercian Fathers Series, 9 (Washington, DC, 1974) pp. 89–90.

45. Bell, *Image and Likeness*, pp. 251–6 and *passim* argues very forcefully for the Augustinian nature of William's thought.

46. See n. 28.

47. To mention two: D. Berger, 'The Attitude of St. Bernard of Clairvaux toward the Jews', *Proceedings of the American Academy for Jewish Research*, **40** (1972) 89–108; J. Cohen, '"Witnesses of our redemption", in B. S. Albert et al. (eds), *Medieval Studies in Honour of Avrom Saltman*, pp. 67–81.

48. *De Diligendo Dei*, 2–3, ed. and trans. E. G. Gardner (London [1916]) pp. 38–45.

49. *De Diligendo Dei*, 8–9, pp. 87–97. For a full discussion of the dating of the *De Diligendo Dei* see *On Loving God with an Analytical Commentary* by E. Stiegman, Cistercian Fathers Series, 13B (Kalamazoo, MI, 1995) pp. 59–66. Stiegman approaches the text in a different way than I do.

50. Berger lists a number of Bernard's anti-Jewish pronouncements in his article, p. 96: 'their "viperous venom" in hating Christ, the bestial stupidity and miserable blindness which caused them to lay "impious hands upon the Lord of Glory" ', their ' "dishonourable [and] serious serfdom" '.

51. Berger, The Attitude at St. Bernard', p. 95.

52. Cohen, 'Witnesses', pp. 67–81.

53. Bernard, *Epistola*, 363, J. Leclercq and H. Rochais (eds), *Sancti Bernardi opera*, 8 (Rome, 1977) pp. 311–17. The question of interest on crusading loans should be linked to the regulations Pope Eugenius had pronounced *vis-à-vis* Christian lenders in the bull *Quantum praedecessores* (1145).

54. I am following the account of the affair by M. Stroll, *The Jewish Pope: Ideology and Policy in the Papal Schism of 1130* (Leiden, 1987) pp. 156–68.

55. See, for example, A. Grabois, 'Le schisme de 1130 et la France', *Revue d'Histoire Ecclésiastique*, **76** (1981) 593–612 and his 'From "Theological" to "Racial" Antisemitism: The Controversy of the Jewish Pope in the Twelfth Century' [in Hebrew], *Zion*, **47** (1982) 1–16.

56. *Epistola* 139, Leclercq and Rochais (eds), *Sancti Bernardi opera*, vol. 7, pp. 335–6.

57. *The Ecclesiastical History of Orderic Vitalis*, ed. and trans. M. Chibnal, vol. 6, Oxford Medieval Texts (Oxford, 1978) pp. 266–9; see also her *The World of Orderic Vitalis* (Oxford, 1984) pp. 158–61.

58. Stroll, *The Jewish Pope*, p. 161; *Arnulphi Sagiensis . . . Invectiva in Girardum Engolismensem Episcopum*, ed. J. Dieterich, MGH, Libelli de Lite, 3 (Hanover, 1897) pp. 92–6; 107.

59. '*nondum* fermento Iudaicae corruptionis penitus expiata', Dieterich, *Arnulphi*, 107.
60. *Contra perfidiam Judaeorum*, PL 207, col. 821; on Peter and the Jews see my 'Twelfth-century Christian Expectations of Jewish Conversion: A Case Study of Peter of Blois', *Ashkenas*, **8** (1998) 1–26.
61. *Epistola* 21 in PL 211, col. 633; letter 38 in J. Bouvet (ed. and trans.), *Correspondence d'Adam, Abbé de Perseigne*, Le Mans: Société historique de la province du Maine, Archives historiques du Maine, 13, 1952–62, p. 352. See also P. Máthé, 'Innerkirchliche Kritik an Verfolgungen im Zusammenhang mit den Kreuzzügen und dem Schwarzen Tod', in S. Lauer (ed.), *Kritik und Gegenkritik in Christentum und Judentum* (Bern, 1981) p. 95.
62. *Adam de Perseigne, Lettres*, ed. and trans. J. Bouvet, Sources Chrétiennes, 66 (Paris, 1960) pp. 7–29; Bouvet, *Correspondence d'Adam*, p. 12; A. J. Andrea, 'Adam of Perseigne and the Fourth Crusade', *Cîteaux Commentarii Cistercienses*, **36** (1985) p. 37; R. Chazan, *Medieval Jewry in Northern France: A Political and Social History* (Baltimore, MD, and London, 1973) pp. 74–5.
63. PL 211, coll. 653–9; Bouvet, *Correspondence d'Adam*, pp. 352–63.
64. Bouvet, *Correspondence d'Adam* pp. 12–16, 111–17, Andrea, 'Adam of Perseigne', pp. 34–7.
65. Bouvet writes that the sanctification of the clergy was perhaps Adam's most significant preoccupation (*Correspondence d'Adam*, p. 112). See also Lotter, 'The Position of the Jews', pp. 182–5.
66. McGinn, *The Calabrian Abbot*, pp. 18–30.
67. E. Pásztor, 'Joachim v. Fiore', *Lexikon des Mittelalters*, 5/3 (Munich and Zurich, 1990) col. 486.
68. *Adversus Iudeos di Gioacchino da Fiore*, ed. A. Frugoni (Rome, 1957) p. 3.
69. See on the Jews of southern Italy D. Abulafia, 'Il Mezzogiorno peninsulare dai bizantini all'espulsione (1541)', in *Storia d'Italia, Annali*, 11, 2 parts, *Gli Ebrei in Italia*, ed. C. Vivanti (Turin, 1996) part 1, pp. 5–44.
70. B. Hirsch-Reich, 'Joachim von Fiore und das Judentum', in *Judentum im Mittelalter: Beiträge zum Christlich-Jüdischen Gespräch*, ed. P. Wilpert (Berlin, 1966) pp. 228–63; H. Schreckenberg, *Die Christlichen Adversus-Judaeos-Texte (11.–13. Jh.). Mit einer Ikonographie des Judenthemas bis zum 4. Laterankonzil*, vol. 2 (Frankfurt am Main, 1991) pp. 346–7, does not see it as a missionizing piece.
71. Frugoni, *Adversus Iudeos*, p. 101.
72. Ibid., p. 9.
73. Ibid., p. 23, cf. p. 29.
74. Ibid., p. 48; Schreckenberg, *Die Christlichen Adversus-Judaeos-Texte*, p. 352.
75. Hirsch-Reich, 'Joachim von fiore', *passim*. There is a vast literature on Joachim's prophecies. McGinn gives a useful selection in his *The Calabrian Abbot*, including the essential studies by Marjorie Reeves.
76. McGinn, *The Calabrian Abbot*, pp. 126–7.
77. *Expositio magni prophete Abbatis Joachim in Apocalypsum...* (Venice, 1527) f. 95ra; trans. McGinn, p. 127 (square brackets are mine).
78. *Expositio*, f. 36vb; Hirsch-Reich, 'Joachim von Fiore', pp. 229–336; idem, 'Die Quelle der Trinitätskreise von Joachim von Fiore und Dante', *Sophia*, **22** (1954) 170–8.

79. McGinn, *The Calabrian Abbot*, pp. 27, 36, 116, 167–8; Hirsch-Reich, 'Joachim von Fiore', pp. 237–43, where she argues that there is no evidence for supposing Jewish origins for Joachim (p. 363).

80. On Joachim's influence see M. Reeves, *The Influence of Prophecy in the Later Middle Ages: A Study in Joachism* (Oxford, 1969); see also under n. 75.

5

'They tell lies: you ate the man': Jewish Reactions to Ritual Murder Accusations

Israel Jacob Yuval

Judel Rosenberg was one of the most popular Jewish writers in eastern Europe at the beginning of the twentieth century. In his book *Nifla'ot Maharal* (*The Miraculous Deeds of Rabbi Loew*, 1909), he described the legend of the creation of the *golem* by Rabbi Judah Loew, the well-known Rabbi of Prague (d. 1609). Rosenberg's book, in which the *golem*'s role was to save the Jews from the terrible outcome of the blood libels, was the first to link the *golem* with blood libels. The book became a great success, indeed a best-seller.

Gershom Scholem traced the historical origins of the story's widespread circulation to the pogroms that dominated the life of eastern European Jewry at the end of the nineteenth century.[1] Eli Yassif noted that the motif of the *golem*'s immediate revenge – on the non-Jew who attacks the Jew – first appears in Rosenberg's story.[2] Scholem and Yassif concur in their view of the *golem* in Rosenberg's story as an historical response to external pressure. Moshe Idel, however, does not accept that the story is only historically motivated. He sees it also as an immanent development that stems from the accumulation of legends about the *golem* throughout the generations.[3]

Nonetheless, Scholem's historical explanation is well supported by the words that Rosenberg attributes to R. Loew when he appeals to the *golem* whom he has just created: 'Know that we created you from the dust of the earth in order to protect the Jews from all evil and from all the sorrows that they have suffered at the hands of their enemies... and you must listen to my voice and do everything I command you.' Thus, the *golem* is an all-powerful creation who can protect the Jews, but is controlled by the rabbi. The book describes three incidents in which

R. Loew employs the *golem* to save the Jews of sixteenth-century Prague from blood libels. And while the main role of the *golem* is to seek out and expose the libel, he is sometimes made to punish the perpetrators too.

It may well be that the great impact that Rosenberg's story had – from its time of writing at the beginning of the century right up to its acceptance as an undisputed historical fact among ultra-Orthodox Jews of modern-day Israel – lies in the fact that it provides traditional Jewry with an answer to the challenge posed by secular Zionism. The story of the *golem* offers an alternative solution that does not necessitate a Jewish exodus from Europe but instead allows the community's traditional leadership to maintain its authority. Since the power wielded by the *golem* is controlled by the traditional leadership (i.e. R. Loew), the story of the *golem* counters the story of Zionism. In his stories of the blood libels of sixteenth-century Prague, Rosenberg offers an alternative historiography that provides a new response to Jewish suffering.

The legend of the *golem* of Prague thus offers a Jewish response to the blood libels, a response that does not depend, as might be expected, solely on the denial of a fraudulent accusation, but also exposes the lie and punishes those who spread it.[4] While this explanation accounts for the modern Jewish response to Christian violence, can it also be applied to the Middle Ages, the period when the blood libel made its debut on the historical stage? Did the Jews of those times take active steps to deny and fight it?

The *Ma'aseh Nissim* (*Miraculous Deeds*), written in Worms in the seventeenth century, describes an event that appears to attest to a Jewish readiness to meet force with force.[5] According to the account, at the time of the Black Death in 1349, when the Jews of the city saw that they were being accused by the town council of poisoning the wells and were about to be sentenced to death, the community's leaders – 12 in all – decided to respond with violence. They appeared for trial before the town council with small daggers hidden beneath their clothes. When the verdict was given, the leaders attacked the council members and killed them. A similar episode is recounted in *Shevet Yehudah* (*Sceptre of Judah*) by Salomon Ibn Verga, written in the first half of the sixteenth century.[6]

While one can surely assume that the incident depicted in this book did not actually take place, it does reflect how Jews of the early modern period wished to see themselves. Jews in the Middle Ages were reluctant to use violence in acts of revenge, holding that revenge should be deferred until the messianic age. When recalling the names of those Jews who had died at the hands of a non-Jew, Ashkenazic Jews would commonly add: 'God will revenge his blood': God will take revenge, not man. As I have discussed elsewhere, the idea of revenge was an essential element in the messianic scenario of the Jews of Ashkenaz.[7] The outburst of vengeance displayed against the members of the town council in Worms reflects the repressed hopes of the Jews of the early modern period rather than those of the Middle Ages.

As shall be shown later, this does not mean that Jews in the Middle Ages did not undertake practical initiatives to refute the libels voiced against them. What is surprising, though, is that there are minimal reverberations of these Christian accusations in Jewish medieval sources.[8] Evidence exists of approximately 90 blood libels until 1600, and yet Jewish literature tends to ignore them. Despite the dark shadow cast over Jewish life by the numerous ritual murder accusations, the Jewish response seems pale and flaccid. While a denial is voiced here and there, almost as an afterthought, Jewish writers fail to employ the moral, philosophical and rhetorical tools in order to prove the futility of the false rumours about their religious customs. Indeed, if vigorous rejections of the accusations do exist, then they come from Christian kings or popes. One can assume that the Jews knew that a firm response on their part would not be effective enough and therefore they invested their efforts in pleading their innocence in royal and papal courts. Only a few cases indicate that this type of solicitation took place, but it can be assumed that it existed in many other cases. In this chapter, I shall try to describe some characteristics of the Jewish reaction to ritual murder accusations in the Middle Ages, in particular in Germany and France.

Evidence of diplomatic and political action taken to refute such charges first appears shortly after these accusations were first made. Indeed, in 1171 the Jewish leadership appeared before King Louis VII's court in order to acquit, *post factum*, the Jews of Blois who had been convicted of drowning a Christian boy in the Loire on Maundy Thursday.

Louis VII was the first of the European kings to grant the Jews a written decree absolving them of all blame: 'Even if the gentiles find a murdered gentile in the town or its vicinity, I will not lay any blame on the Jews, therefore do not be concerned about that.'[9] The King's declaration was accompanied by a promise to ensure the welfare and safety of all Jews in his country. Count Henry of Champagne, the brother of Theobald of Blois who had ordered the Jews in the city to be burned to death, also declared: 'we have found no evidence in the Jew's Bible that it is permissible to kill non-Jews.'[10] The strong reaction of the ruling authorities appears to be based not only on the Jews' denial of their guilt but also on a prevalent scepticism regarding ritual murder libels when they made their debut in the twelfth century. Furthermore, in the Blois case, no corpse was found, a fact that greatly weakened the Christian claim. In 1187 too, the Bishop of Mainz was appeased by the Jews' promise that they had not laid a finger on a Christian who had falsely accused them of wanting to kill him. The bishop asked the Jews to swear that 'they do not kill any non-Jews on the eve of Passover'.[11] Their subsequent declaration sufficed and the suspicions were appeased.

By contrast, in 1236, in the wake of the Fulda blood libel, a change took place. After being presented with the corpses of the dead children, Emperor Frederick II ordered an inquiry. At first, he assembled princes, magnates and nobles of the empire, as well as abbots and ecclesiastics, to question them about the accusations against the Jews. Opinions were divided, and there was no consensus. Only after converts from western countries outside the German Empire were summoned to the emperor's court, and gave testimonies that the claims were false, were the Jews completely exonerated. Thus, like his predecessor the French King Louis VII, in the summer of 1236 Emperor Frederick II refuted the accusations and vindicated the Jews of all charges. In this case, however, the acquittal was based not only on the Jews' own declaration of their innocence, as in the past, but also on the testimony of a third party, converts from beyond the Empire. The Jews themselves were by now considered untrustworthy, and suspicions were growing that their religious rituals might include criminal deeds hitherto hidden from the Christians.

Could this lack of trust have emanated from a new motif first seen in the Fulda blood libel, which transformed the Jews into a demonic and far more dangerous enemy? It is commonly accepted by researchers that the Fulda blood libel was the first in which the former charge of ritual murder by means of crucifying a Christian child was replaced by a charge of ritual cannibalism and murder in order to obtain the victim's blood.[12] Gavin Langmuir claimed that until the Fulda libel in 1235, not a single ritual murder libel similar to those in thirteenth-century England and France had occurred in Germany. Thus he sought and found – in Conrad von Marburg's strident sermons against heresy – a local factor that gave rise to the Fulda blood libel of 1235.[13]

As mentioned previously, however, the demand made of the Jews of Mainz – to swear that they did not kill any non-Jews on Passover eve – shows that by 1187 there already existed suspicions in Germany that Jews did, in fact, commit ritual murder. Indeed, if there was no suspicion that ritually motivated murders were being committed, why then was suspicion raised in specific connection with Passover eve?[14] Furthermore, the claim that the motif of blood first appears in 1235 is also questionable in light of two sources that seem to attest to an earlier appearance of the motif of cannibalism. In his chronicle, William the Breton presents an item from Rigord's chronicle, according to which, in 1180, the Jews of France were being accused (apparently on the basis of the incident in Blois in 1171) of killing a Christian every year and eating his heart.[15] Further doubt is cast by the Hebrew *piyut* (liturgical poem) 'Elohim Hayyim' (Living God), written by Rabbi Shlomo ben Avraham to commemorate the persecution in Erfurt in 1221. The *piyut* includes the following:[16]

> They tell lies: you ate the man
> They ate me the flesh and the blood
> And only to them the land was given
> 'If it pleases the king, let a decree be issued to destroy them'.[17]

One could interpret these lines as referring to the accusation that the Jews crucified Jesus. The poet describes the Christian claim that the Jews 'ate', i.e. killed, 'the man', i.e. Jesus. These lines are based on Ezekiel 36: 13: 'Thus said the Lord, because they say to you, you eat man, and bereave your nations.'[18] In the original, 'you eat man' appears without the definite article, but in the *piyut* it appears as 'you eat the man', with the definite article. This change could be interpreted as stressing the human nature of Jesus, the Man, and refuting his divine nature as the Son of God.

In light of this, the meaning of the Jewish response in the *piyut* is understandable: 'They ate me the flesh and the blood'. The suffering and affliction of the Jews under the Christians turned them, not Jesus, into a flesh-and-blood sacrifice. As Andreas Angerstorfer (p. 140) pointed out, the repeated use of the verb 'eat' in this sentence can be interpreted as reversing the Christian claim. The use too of the 'flesh and blood' motif reinforces the impression that the Jewish suffering is presented as a counter-parallel to the Eucharist. Similarly, the next sentence 'And only to them the land was given' can be interpreted as a continuation of the description of the Christians' ill treatment. They 'eat' the Jews, i.e. kill them, and claim that only Christians have the right to live in the Christian land. Such an attitude completely refutes any basis for Jewish existence within Christian society, and is thus in keeping with the new tendencies of the thirteenth century as observed and understood by the Jewish *piyut*.[19] Thus, in this analysis, the *piyut* is not referring to the cannibalistic motif of blood libels.

An alternative interpretation of the poem is based on a similar passage in *Sefer Nizahon Yashan* (also known as *Sefer Nizahon Vetus*, *Old Book of Contention*, to distinguish it from Rabbi Yom Tov Lipmann Mühlhausen's *Sefer Nizahon* – see below), an anti-Christian polemic from the end of the thirteenth century.[20] The parallels between the two texts are remarkable:

Sefer Nizahon Yashan	*Elohim Hayyim*
They shall eat it with unleavened bread and bitter herbs (Numbers 9: 11):	
Refers to the fact that the nations subject us to hard work, embitter our lives, and revile us by saying that we eat human beings and the blood of Christian children. Indeed, the heretic may attempt to bolster this last assertion by arguing that	They tell lies: you ate the man
Ezekiel refers to such a practice when he says to the Land of Israel, 'Thus said the Lord, Because they say unto you, You eat men, and bereave your nations . . . ' (Ezekiel 36: 13). The people of Israel are also called 'land', as it is written, 'You shall be a delightful land' (Malachi 3: 12); thus, the verse in Ezekiel must refer to the people of Israel.	And only to them the land was given
	They ate me the flesh and the blood
The answer to this argument is that Scripture can be cited to prove that they too eat human beings, as it is written, 'For they have eaten up Jacob' (Jeremiah 10: 25).	

Quite clearly, *Sefer Nizahon Yashan* is not discussing the crucifixion of Christ but rather the charge of cannibalism against the Jews. The accusation that the house of Israel are consumers of blood is given as an actualization of the verse 'They shall eat it with unleavened bread and bitter herbs (Numbers 9: 11)', thus reinforcing the impression that the Christian accusations referred to the Jews' purported use of Christian blood to bake unleavened bread.[21] Then comes an entry which presents a Christian homily on Ezekiel 36 according to which Jews eat human beings.[22] But since the passage in Ezekiel does not discuss the people of Israel but rather the Land of Israel as devouring its inhabitants, a passage from Malachi 3: 12 is integrated in order to show that the people of

Israel and its land are identical. Thus, just as the land consumes its inhabitants, so too do the Jews.

The Jewish response to this Christian claim is surprisingly feeble. It contains no rebuttal of the Christian accusation, just a counter-accusation: you too are cannibals. The passage from Jeremiah was selected since the verb 'eat' is used in it – the non-Jews eat Jacob – and thus it provides a counter-response to the Christian claim which uses Ezekiel 36 to show that the people of Israel are cannibals.[23] There could be another purpose for quoting Jeremiah. The verse in full is: 'Pour out your wrath on the nations that do not acknowledge you, on the peoples who do not call on your name. For they have eaten him completely and destroyed his homeland'. These verses gained fame when, during the Middle Ages, they were incorporated into the Passover Haggadah as the opening declaration for the pouring of the fourth cup, a symbol of the last redemption from the Edomite kingdom.[24] The inclusion of 'Pour out your wrath' in the Passover Haggadah can thus be considered an internal Jewish response to the ritual murder libels of the Middle Ages.

The argument in *Sefer Nizahon Yashan* is over the significance of the eating of matza and bitter herbs, as mentioned in the verse at the beginning of the passage. The Christians charged the Jews that the unleavened bread eaten by them was baked with Christian blood, and thus symbolized Jewish murderousness and Christian martyrdom at the same time. The Jews responded that they were persecuted and killed, and therefore the unleavened bread symbolized both Jewish martyrdom and Christian murderousness. It should be noted that the very same motif appears in another anonymous *piyut* about a blood libel in Munich in 1285.[25]

> An impure land, the city of priests
> Cooked with unleavened bread and bitter herbs.

The poet describes the blood libel in Munich ('city of priests'), and the disastrous fate of the Jews who were burned to death ('cooked'). The phrase 'cooked with unleavened bread and bitter herbs' is the equivalent of the verse 'they shall eat it with unleavened bread and bitter herbs' (Numbers 9: 11), so that 'cooking' substitutes 'eating'. Here too the unleavened bread and bitter herbs reflect the suffering of the Jew who is cooked and consumed by fire. As the poem continues, so too the word 'eating' is used to symbolize annihilation and killing:

> So said the burnt: 'You ate the rest of my people
> Therefore you will be eaten by the fire of the redeemers'.

Those speaking are God's martyrs who have been burned ('the burnt'), and they are addressing those who have set them alight, who have eaten the people's remains, prophesying that, in retaliation, they will be consumed by a fire of revenge when the time of redemption comes.[26]

As mentioned above, the thematic similarity between the *piyut* 'Elohim Hayyim' and the *Sefer Nizahon Yashan* is evident. A parallel reading of the two reinforces the sense that the *piyut* was also trying to confront Christian claims of Jewish cannibalism. Both passages are consistent in their use of the verb 'eat' as a synonym for murder, and both make direct or implied use of Ezekiel 36: 13. Both passages describe eating 'the man' with the definite article, even though the verse in Ezekiel does not use the definite article.[27] Indeed, in the same way, one can explain the response to the non-Jews in the *piyut* 'Elohim Hayyim' ('and only to them the land was given') as countering the Christian claim: the land that you say to be identical with Israel is not yours. The *piyut* was written during crusader rule of the Holy Land, when the people of Israel and its land were no longer identical.

This interpretation of the poem 'Elohim Hayyim' pre-dates the appearance of the cannibalistic motif in the Fulda events of 1235, and casts doubt on the firm distinction that scholars tend to draw between ritual murder libels and blood libels, as if the two were entirely different phenomena. In truth, ritual murder libels and blood libels are both associated with the same season – Passover and Easter. The ritual murder accusations associated Jewish murderousness with Easter and the crucifixion, while blood libels associated it with the Jewish Passover and the eating of unleavened bread. It seems likely that the second motif was a natural and indiscernible development of the first, not only because the two festivals are held at a similar time, sometimes even on the same days, but also because Passover and its Jewish contents are very similar to Easter and its Christian contents. The change from the motif of the crucifixion of a Christian child in order to hasten the Jewish redemption, a motif already apparent in the Norwich libel based on a later interpretation of a murder case that occurred in 1144, to the motif of eating unleavened bread – symbolizing martyrdom and poverty but also the awaiting redemption in the future – is a much less dramatic transformation than might appear. In fact, with regard to the Blois incident, the Hebrew and Latin sources do not mention the motif of cannibalism. Thus, its appearance in Rigord's *Chronicles* shortly afterwards indicates an indiscernible change from ritual murder at Easter to Jewish cannibalism at Passover.

Elsewhere in *Sefer Nizahon Yashan* a slightly more vigorous response to the blood libels is offered.[28] The Jewish responder offers two denials of blood libels. The first, that murdering a non-Jew is forbidden in the Torah ('"Do not murder" refers to any man; thus we were warned against murdering non-Jews as well.'); the second, that Jews avoided eating animal blood, thus inferring they would avoid human blood. Why then would the Christians make these false accusations? 'You are concocting allegations against us in order to permit our murder.' Again we see the trend to overturn the Christian claim. Thus, the author counters the 'Jewish' blood libel with a 'Christian' blood libel: the Christians want to set Jewish blood flowing so they falsely accuse them of requiring Christian blood. Indeed, the Jew on his part does not claim that the Christian needs blood for his ritual acts (a claim made in another source which will be discussed later), but he provides a literary parallel of blood for blood.

These sources attest to a meagre reservoir of Jewish responses to the allegations raised against them. This is illustrated clearly in an episode in *Sefer Joseph Hamekane* (Book of Joseph the Zealot, France, thirteenth century), recounting a conversation in Paris between the 'Chancellor' and two rabbis.[29] The Christian claimed that from the verse 'and you shall drink the blood of the slain' (Numbers 23: 24) it could be deduced that 'you eat blood of the uncircumcised', to which the two Jewish responders 'stood and did not reply'. The author criticizes their inability to give an appropriate answer.

This Christian interpretation of 'and you shall drink the blood of the slain' is also cited by Rabbi Yom Tov Lipmann Mühlhausen in his *Sefer Nizahon* (*Book of Contention*) from approximately 1400. But he also cited Numbers 24: 8 ('he shall eat up the nations his enemies') and Deuteronomy 7: 16 ('and thou shall consume all the peoples').[30] All these verses, says Lipmann, were used by the Christians in their claim that the Bible and its promises of the future are the source of the use of blood by Jews. Like his predecessors, Lipmann also preferred to present an alternative Jewish interpretation of these verses rather than deny the accusation itself. The verb 'eat', he claims, has no connotation of killing non-Jews, but rather of conquering and suppressing them. He interprets the verse 'you shall drink the blood of the slain' as meaning: in war, the Children of Israel will beat the bad ('slain') and will take their money (*damim* in Hebrew means both blood and money) and use it for feasting ('will drink'). This is a very twisted interpretation, and it indicates a limited understanding of the severity and danger that was inherent in the Christian accusations regarding the blood libels. Lipmann thus perpetuated an apologetic tradition that suffices with rejecting the enemy's claims without trying to counter-attack.

These Jewish responses, few in number and weak in content, are confined to biblical-interpretative language which fails to perceive the need to present practical arguments – legal and ethical – to prove the foolishness of the libel. Jewish apologetics were thus driven by the Christian claims. The Jewish reaction tended to confront the Christians with a counter-claim rather than focus on the denial itself, and this too was because it came on the heels of the Christian initiative. The argument is in the form of traditional exegesis. The Jews want to refute the Christian interpretation of the biblical verses to prove Jewish murderousness. The Jews present a counter-interpretation.

Another characteristic that is common to all the sources presented here is the exclusive focus on clarifying the meaning of the word 'eat' in the Bible. The Christians interpreted it as meaning the killing of non-Jews by Jews. The Jews tended to concur with this interpretation of the word 'eat' although they presented other verses in which those killed are shown to be Jews. Why was the entire argument focused on the meaning in the Bible of the word 'to eat' and on who ate whom?

The connection between eating and killing stands in contradiction to the prevailing attitude in the Middle Ages whereby eating was perceived as a means of communing with God. Starvation and thirst symbolized longing for God and for Christ, and the ability to touch his lips. At Communion worshippers participate in eating the Son-God, thus identifying with him in a mystical union.[31] Abstaining from food also helps one get close to God, thus providing the connection between fasting and death and martyrdom. Abstinence from terrestrial eating is regarded as preparation for receiving the Eucharist – the divine bread – since it is identified with Christ's Passion and is a realization of the ideal of *Imitatio Christi*.[32] Someone who fasts is comparable to a voluntary martyr who is entitled to the celestial bread instead of the bread of the earth, thus turning the eating of the sacrament into a revelation of the grace of God.

However, the Hebrew root of 'eat' has a double meaning of both 'eating' and 'annihilation' that is common in the Bible[33] and in Talmudic language,[34] and frequently appeared in Ashkenazic poetry before the thirteenth century.[35] In the Hebrew chronicles of the First Crusade the deaths of the martyrs are given a comparative meaning of sacrifice, thus turning them into a 'food' that was offered on the heavenly altar. Although the close connection between these two meanings was by no means an innovation of the thirteenth century, an entirely different association was created in the almost exclusive use of the motif of eating in the Jewish polemic against the blood libel that took place within a Christian environment. In this context, eating the host and drinking the holy wine acquired the meaning of martyrdom. The ritual murder libel or blood libel is based on the victim being perceived as a sacrifice. The murderer 'eats', the sacrifice is 'eaten'. The Jewish language internalized the Christian notion of the Eucharist as it is expressed in the eating of the host by the congregation and the drinking of the wine by the priest.

This process of internalization is also evident in a bull issued by Pope Innocent III in 1205 and addressed to the archbishop of Sens and to the bishop of Paris, demanding that they forbid Jews from employing Christian wet-nurses. He claimed that, during the three days following the Lord's Resurrection (Easter), when the wet-nurses received the host, the Jews pour their milk into the latrine so that their children would not drink milk derived from the host.[36] The Pope regarded such behaviour as both impudent and a denigration of the Christian faith. Baron describes such accusations as 'unverified rumours ... [that] can be understood only against the background of the widespread folkloristic accusation of Jews' desecrating the host, a practice which, in the popular imagination, was almost invariably followed by miracles'.[37] According to Baron, the Pope's words attest to a Christian figment of imagination, rather than a Jewish practice. It seems though that Baron has momentarily forgotten that, as far as we know, the first accusation of desecration of the host was raised in Paris in 1290, 85 years after Pope Innocent's letter!

Indeed, it seems that there is truth in the Pope's claims regarding Jewish customs in his day. In the Talmud there is discussion as to whether one is permitted to employ a gentile wet-nurse in order to suckle a Jewish baby.[38] The point in question was whether a non-Jewish wet-nurse might endanger the life of a Jewish baby. Both sides agreed that the type of food – kosher or non-kosher – eaten by the wet-nurse played no role. In Ashkenaz, the opinion allowing the employment of wet-nurses was accepted, and the custom of employing Christian wet-nurses was widespread with no restriction on what they ate.

However, in the second half of the thirteenth century, this situation changed. Rabbi Yitzhak ben Moshe 'Or Zarua' (1180–1250) was the first to impose restrictions on food eaten by the Christian wet-nurse: 'The [Christian] wet-nurses should be warned not to eat unkosher food and pork, certainly not unclean things'.[39] It is more than likely that 'unclean things' are foods that are considered idolatrous, i.e. the host. The proximity in time between the pope's claim and Rabbi Yitzhak Or Zarua's new ruling proves that a new practice had indeed become current among Jews at the beginning of the thirteenth century. This practice shows a growing Jewish sensitivity to the significance of unclean foods in general, and in particular of the host. Indeed, the pope's testimony that Jews would pour the wet-nurses' milk down the latrine can be interpreted as a Jewish custom based on the Babylonian Talmud, Gittin 57a, where Christ's punishment in the next world is to be put in 'boiling excrement'. Christian writers of the thirteenth century were familiar with this Talmudic verdict, and its words were perfectly in keeping with the motif in later host-desecration accusations that Jews threw the stolen and desecrated host down the latrine.[40]

The pope's testimony, however, pre-dates by as much as a generation the new ruling of R. Yitzhak Or Zarua, and thus his prohibition must have been the result of an instinctive recoiling that only later gained legal authorization. This new Jewish sensitivity was, it can be assumed, a reaction to the intensive theological debate among Christians regarding the significance of the host. The debate came to an end in 1215 when Transubstantiation became dogma. At the same time, the Jewish practice reflects the increasing significance of food in the sacred language that was common to both faiths.

Ivan Marcus has recently drawn a comparison between school initiation rituals for Jewish children in the Middle Ages and similar First Communion ceremonies among Christians: 'Both required that the child ingest symbolic foods among which a special kind of ritually sanctified cake or bread had a prominent place'.[41] In particular, Marcus points to the similarity between eating cakes baked in the form of Hebrew letters and the first eating of the host. The Jewish use of Christian symbols aimed to deny the Eucharistic devotion and to claim that the Torah with its sacred letters is the true bread and the *corpus Dei* (body of God), not the host which is the *corpus Christi* (body of Christ).

Israel Ta-Shema has rejected this comparative presentation of Jewish and Christian symbolic language.[42] In his opinion, the comparison is merely at the level of appearance, and the Jews had in no sense culturally internalized Christian symbolism: 'One cannot totally deny the possibility that the external, visual similarity between the two [sets of] symbols inspired a few individuals among the two religions to amuse themselves with this kind of ideological hair-splitting.'[43] Ta-Shema's marginalization and minimization of the parallel phenomena are evident in almost every word of this sentence.

Nevertheless, the comparison between Innocent III's bull and R. Yitzhak Or Zarua's legal ruling supports Marcus's claim, since it shows that the Christian language infiltrated far beyond the peripheries of the Jewish populace. Indeed, a prohibition which emanated from a religious sentiment became the legal ruling of a renowned Halachist. Furthermore, Marcus is discussing Christian symbolism that during a period of 'acculturation' amongst Jews gained internal 'Jewish' meaning. The episodes that have been discussed here point to a different level of internalization, whereby Christian symbols gained an ontological status – albeit in a negative sense – even in the eyes of the Jews. The Christian symbol was transformed into a weapon that the Jews used in retaliation. There was no rebuttal of the symbol's credibility or power, but rather an attempt to expropriate its sacred status by counter-attacking, as in the case of pouring the milk into the latrine. This too was the nature of the language used by the Jews to confront the blood libels, a language not so much of denial but rather one of reversal.

Another *piyut* on the Munich blood libel of 1285, by Rabbi Hayyim ben Machir, illustrates this well.[44] The poem is structured as a dialogue between Christian deeds and claims poised against the Jews' response. The beginning of the *piyut* describes the ruler's ruse to expel the Jews from the city and take over their property. He is aware that the people suspect the Jews of killing Christian children, of destroying their limbs and spilling their blood and drinking it. Thus, the ruler secretly kills a Christian boy, dismembers his corpse and presents it to the masses. The crowd immediately becomes agitated and besieges the synagogue where the Jews have gathered for their Sabbath eve prayers. The Jewish reaction to this accusation is an immediate and outright denial: 'Let it be known to everybody that we do not spill the blood of Christian children.' This type of 'sacrifice', they declare, is regarded by the Bible as 'impure and [it] will not be accepted (Leviticus 19: 7)'. Then the boy's corpse was bought before them and they were accused of murdering him, but the Jews retorted: 'They screamed in bitter voices: We are innocent, take our possessions but spare us our lives. Why would you spill innocent blood?' Those accusing the Jews demanded that they be punished or forced to convert to Christianity. The Jews refused to convert and 90 were burned to death.

The *piyut* is surprising in its use of the motif of sacrifice. Immediately after describing the Christian claim that the Jews slaughtered Christian children so that they could drink their blood, the *piyut* describes the Jews who have been killed in terminology directly taken from the arena of sacrifices: 'They were all burnt as an offering with an aroma pleasing to the Lord; the fat parts which were meant to be sacrificed during the week were allowed to be sacrificed on the Sabbath ... the limbs of the father and mother, son and daughter substituted the *tamid* (perpetual offering) and the *mussaf* (additional offering.) Males and females were all accepted as a burnt offering.' The Jewish apologetic language – albeit unconsciously – is semantically akin to that used by the Christians in their accusations.

It would seem that this is a paradox common to many Jewish reactions to blood libels during the Middle Ages. For historians of the third millennium, the dialogue between the Jews and Christians appears very similar. In fact, each side was immersed in an internal dialogue with itself though employing the very same language that the other was using. The Jewish denial of blood libels took the form of a counter-claim. While the blood libel was an attempt to depict the Jew as murderer and the Christian as victim, the Jewish response sought to reverse these roles and to restore the role of victim to the Jew.[45] In doing so, the Jews used concepts that were familiar to both their own world and that of Christian martyrdom. Thus, while the human sacrifice is eaten or drunk in a manner that suggests the eating of the host and the drinking of wine in the Eucharist, the description of the Jews being burnt comes directly from the world of the temple offerings on the altar.

This proximity can perhaps better explain what appears to be Jewish 'silence' *vis-à-vis* blood libels in the Middle Ages. The Jewish reaction to blood libels appears less articulated than we might have expected, because although it denied any Jewish culpability, it internalized some of the narratives which nourished the ritual murder accusations. Many blood libels led to the emergence of Christian hagiographical literature, whose aim was to justify the creation of a new saint or a new site of pilgrimage. This is what happened in Norwich in the twelfth century and in Lincoln and Bacharach in the thirteenth century. Likewise, Jews also created their own accounts of Jewish martyrs who died for the Sanctification of the Holy Name. Thus, a certain aspect of the blood libels – the motif of martyrological sacrifice – was not alien to the Jews in the Middle Ages. This internalization led to Jewish counter-narratives, whose motifs were similar to its Christian equivalent but whose meaning was entirely polemical.

An example of this internalization is seen in a Jewish apocalyptic text written in the twelfth or the thirteenth century, known as *Rabbi Shimon ben Yochai's Prayer*.[46] The text describes an eschatological Christian–Muslim war at whose end '[the Christians] removed the brains of the [Muslim] babies, and slaughtered babies to Christ every day'. It seems that this is an attempt to accuse the Christians of the very same act that they accused the Jews: namely, the ritual murder of babies.

The murder of infants is, of course, a recurrent motif in world literature. The murder by Pharaoh of the male offspring of the Israelites in Egypt is an archetypal tale that is also the source of the Christian story of the slaughter of the Innocents in Bethlehem, on the command of Herod. In the fourth century, it led to the legend of Sylvester. The Emperor Constantine had leprosy, and no treatment could be found save bathing in a bath of children's blood. The emperor's soldiers immediately rounded up innocent children, but when the emperor heard the cries of their mothers he had mercy on them. During the night, Peter and Paul appeared to the emperor in a dream and he was cured – leading to his conversion to Christianity. In return for his recovery the emperor gave Sylvester, the bishop of Rome, the famous gift of the Donation of Constantine.[47]

This Christian legend has a Jewish equivalent, in which Pharaoh replaces Constantine and the children destined to be killed are Jewish children. In the Jewish story too, there is a miracle at the last minute, and they too are saved. This version underwent a further development at the end of the Middle Ages: in illustrations in the Passover Haggadah, the miracle disappears. Rather, there are violent and brutal depictions of Pharaoh's soldiers murdering the children and of their mothers' pleading. David Malkiel rightly interprets this change as indicating a Jewish response to ritual blood libels.[48] It is an attempt to claim that, like other children, the children of Jews were slaughtered too. In the symbolic language of the Middle Ages, one can also see here a form of protest: since Pharaoh was the archetypal ruler who symbolized the enslavement of the Children of Israel, so too the babies of the Children of Israel were identifiable with Jewish children in later ages.

Thus, the Jews did not only refute the libels against them but they also internalized them and even fashioned their own counter-stories of ritual murder whose elements fitted their own needs. While these two trends – of denial and of internalization – probably seem contradictory, it must be remembered that there was no unified Jewish attitude on the issue. Alongside the rational denial of blood libels, there was also an infiltration within the Jewish minority of the religious values of the Christian majority, which attributed ritual significance to the blood of martyrs. Such a duality was also present in Christian society in its conflict between the repeated and recurring denial of the accusations against the Jews and the frequent libels that were waged against them.

It is therefore no surprise that the first Jewish work whose main concern was the struggle against ritual murder accusations did not appear until the sixteenth century. This is the book *Shevet Yehuda* of Salomon Ibn

Verga, already mentioned.[49] The author of this book took historical material and turned it into literary fiction in order to reflect his particular view of Jewish history during the period of the exile. In one of these fictitious episodes, he describes a conversation between King Alfonso and the sage Thomas.[50] A certain bishop, who is leading an incitement against the Jews with the claim that they need Christian blood in order to celebrate the festival of Passover, has won the support of the masses. The king decides to investigate the accusations, and asks Thomas to state his position. Ibn Verga is the script writer and producer of this dialogue. He attributes to the king the role of defender of the Jews, albeit a somewhat hesitant one, and Thomas represents the mood of enlightenment, offering a fervent and outright denial of the accusations against the Jews. Thomas presents three arguments. The first argument is that the Jews would not resort to such actions and risk their rather precarious status in the Christian kingdom. The second is that the Jews have a well-known reputation for being non-violent and easily intimidated: 'what [else] can we say about their fearfulness, thus if there are a hundred Jews in the street and a little Christian child comes and says: "Get the Jews!" – they all immediately run away'.[51] The non-violence of the Jews is presented as an inherent characteristic, thus proving the foolishness of believing in ritual blood libels. The third argument is that since Judaism is very strict regarding the prohibition on drinking the blood of animals, Jews could in no way be suspected of consuming human blood.[52]

The king is not easily convinced. He claims that the prohibition of murder, regarded by Jews as a particularly grave crime, pertains only to Jews, not Christians, and thus Jews do not regard Christians as human. The king based his claim on the Talmud, which draws a distinction between the ox of a Jew that kills the ox of a non-Jew and whose owner is exempt from responsibility, and the ox of a Jew who kills the ox of another Jew and whose owner is responsible.[53] Thomas responds that the non-Jew in the Talmud is a pagan, not a Christian.

Whether or not there is a real historical background to Ibn Verga's dialogue, two facts must be emphasized: (a) that the first serious discussion of blood libels in Jewish writing took place only in the sixteenth century; (b) that the discussion is presented as a dialogue between two Christians, with no Jewish participation. It should also be noted that the first serious discussion of the accusations against the Jews in Christian literature also took place in the first half of the sixteenth century. It was the German humanist, Andreas Osiander, who for the first time defended the Jews in his treatise *Ob es war und glaublich sey/dass die*

Juden der Christen kinder heymlich erwürgen/und jr blut gebrauchen (Whether is true or believable that the Jews secretly kill Christian children and use their blood).[54] This book elicited an immediate response by Johann Eck, who defended the veracity of these accusations.

This new discourse is the result of the new Protestant and intellectual approach, which regarded the Catholic dogmas concerning the power of the Mass and the Eucharist as ridiculous.[55] Christian suspicions about the Jewish rituals became in the sixteenth century a matter of public concern, and they required pertinent Jewish responses. This new debate about the veracity of the ritual murder accusations marks therefore the beginning of a new age, in which the symbolic discourse of the Middle Ages was replaced by a more rational tone and a more realistic language.

Notes

1. G. Scholem, *On the Kabbalah and Its Symbolism* (New York, 1965) p. 189, n. 1.
2. Yehudah Yudel Rosenberg, *The Golem of Prague and Other Tales of Wonder*, ed. E. Yassif (Jerusalem, 1991) pp. 50–2. B. L. Knapp, *The Prometheus Syndrome* (New York, 1979) pp. 97–131 offers a psychoanalytical interpretation of this popular story, whereby the historical character of R. Loew acquires the status of a wise old man/elderly sage and even the *golem* is attributed the role of saviour. According to an older version of the story from the year 1893, told by Isaac Leib Peretz (*Alle Werk*, vol. 2 (New York, 1947) pp. 310–11), R. Loew created the *golem* in order to protect the Jews from their persecutors who had attempted to conquer the ghetto.
3. M. Idel, *Golem: Jewish Magical and Mystical Tradition on the Artificial Anthropoid* (Albany, NY, 1990) pp. 255–6 (see a more comprehensive version in Hebrew, Tel Aviv, 1996, pp. 216–17). Idel based his argument on the existence of a Hassidic version of the *golem* legend that was already in print in 1903 and which – he assumed – was in circulation even earlier, before the wave of blood libels began in 1885. In fact, this Hassidic version repeats the story entitled 'der Golam des Rabbi Löw', which had already been published in 1841 by Franz Klutschak in *Panorama des Universums*, 8. See H. J. Kieval, 'Pursuing the Golem of Prague: Jewish Culture and the Invention of a Tradition', *Modern Judaism*, **17** (1997) 11 (the story is reprinted on pp. 21–3). However, the *golem* in this version lacks any of the attributes of a saviour or warrior who defends the Jews in a time of turbulence and persecution.
4. A survey of Jewish responses to ritual murder accusations: H. J. Kieval, 'Talking Back: the Dilemma of Jewish Intervention in the European Discourses of Ritual Murder' (unpublished paper delivered to the Department of the History of the Jewish People at the Hebrew University of Jerusalem, 1997).
5. *Sefer Ma'aseh Nissim* (Amsterdam, 1696) no. 10, fol. 17a–18b. For a real case of intervention by the 12 leaders (*parnassim*) of the Jewish community in Worms

on behalf of an innocent Jew accused in the blood libel of 1563, see R. Po-Chia
Hsia, *The Myth of Ritual Murder: Jews and Magic in Reformation Germany* (New
Haven, CT and London, 1988) p. 189. Their intervention put them in conflict with the magistrates of the city. Another version of this story from 1485
connects this episode with the Black Death in Fulda (1349). See A. David,
'Tales Concerning Persecutions in Medieval Germany' [Hebrew], in Z. Malachi
(ed.), *Papers on Medieval Hebrew Literature Presented to A. M. Habermann on
the Occasion of His 75th Birthday* (Tel Aviv/Jerusalem, 1977) pp. 77–8, 82–3.

6. Salomon Ibn Verga, *Shevet Yehudah*, ed. Y. Baer and A. Shohat (Jerusalem,
    1947), ch. 34, p. 91. For more on the book, see F. Baer, *Untersuchungen über
    Quellen und Komposition des Schebet Jehuda* (Berlin, 1923); Y. H. Yerushalmi,
    The Lisbon Massacre of 1506 and the Royal Image in the 'Shebet Yehudah', Hebrew
    Union College, annual supplement, vol. 1 (Cincinnati, OH 1976).

7. See my forthcoming book *'Two Nations in Your Womb': Perspectives of Jews
    and Christians*, Chapter 3.

8. H. Strack, *Das Blut im Glauben und Aberglauben der Menschheit. Mit besonderer
    Berück-sichtigung der 'Volksmedizin' und des 'jüdischen Blutritus'* (Munich,
    1900) ch. 19, pp. 167–9 lists the Jewish refutations of ritual murder accus-
    ations, beginning with Isaac Abrabanel (d. 1508)! A. Angerstorfer, 'Jüdische
    Reaktionen auf die mittelalterlichen Blutbeschuldigungen vom 13. bis zum
    16. Jahrhundert', in R. Erb (ed.), *Die Legende vom Ritualmord: Zur Geschichte
    der Blutbeschuldigungen gegen Juden* (Berlin, 1993) pp. 133–6 analyses the
    medieval responses.

9. A. Neubauer and M. Stern (eds), *Hebräische Berichte über die Judenverfolgungen
    während der* Kreuzzüge (Berlin, 1892; repr.: Hildesheim, 1997) p. 34; A. M.
    Habermann (ed.), *Sefer Gezerot Ashkenaz ve-Zarfat* (Jerusalem, 1945; repr.
    1971) p. 145

10. Neubauer and Stern, *Hebräische Berichte*, p. 35; Habermann, *Sefer Gezerot*,
    p. 146.

11. Habermann, *Sefer Gezerot*, p. 161.

12. The first to distinguish between the two was Hermann Strack, followed by
    Joshua Trachtenberg, Cecil Roth and lately Gavin Langmuir. For a summary
    of their opinions, see G. Langmuir, 'Ritual Cannibalism', *Toward a Defin-
    ition of Antisemitism* (Berkeley and Los Angeles, CA, 1990) pp. 263–81.

13. Ibid.

14. See n. 11, above.

15. 'Judaei singulis annis unum Christianum immolabant, et ejus corde se com-
    municabant.' See D. Berger (ed. and trans.), *The Jewish–Christian Debate in
    the High Middle Ages: A Critical Edition of the Nizzahon Vetus* (Philadelphia,
    PA, 1979) p. 343. However, Berger claims that the Jews are not accused of
    killing and eating Christians but only that they 'partake of the heart in
    some sort of ceremonial meal'.

16. Habermann, *Sefer Gezerot*, pp. 170–1; A. Berliner, 'Kinot Usselichot', *Sam-
    melband kleiner Beiträge aus Handschriften*, III (Berlin, 1887), pp. 26–7; S.
    Bernfeld, *Sefer Hademaot*, vol. 1 (Berlin, 1924) pp. 266–8.

17. Esther 3: 9.

18. Compare Ezekiel, 19: 3, 6.

19. J. Cohen, 'The Second Disputation of Paris and its Place in Thirteenth-
    Century Jewish–Christian Polemic'[Hebrew], *Tarbiz*, **68** (1999) 567–70, brings

new support to his claim that Catholics in the thirteenth century were more reluctant in following Augustine's policy of tolerance toward the Jews.

20. Berger, *The Jewish–Christian Debate*, no. 16 (with modifications to the translation).
21. Angerstorfer, 'Jüdische Reaktionen', p. 146.
22. For the classical Christian commentary on this verse see Jerome's commentary on Ezekiel, PL 25, col. 340. The literal meaning of the verse refers to the land (compare Numbers 13: 32) and accordingly Rashi commented: 'This land is practiced in destroying its inhabitants; the Emorites were annihilated in it, and the Jews were annihilated in it.' Isaac Abrabanel comments that the verse refers to the people of Israel, and that it is a prophecy of the future blood libels which will occur in exile: 'This alludes to the immense calamity that will happen to us when we are among the Edomites during the exile. They will defame the people of Israel as if they kill non-Jews secretly in order to eat their blood at Passover. This lie and falsity will be the reason for conversions and mass slaughters that the non-Jews will perpetrate upon our people. May God take revenge.' It would be so for the duration of the exile, but after the redemption of Israel, 'they will no longer be the victims of this slander', and there will be no more blood libels.
23. This argument is similar to the one used by a chronographer in Erfurt, the same city where 'Elohim Hayyim' was composed, who utilized Ezekiel 35: 6 in order to justify the massacre of 18 Jews who were accused of murdering a Christian in the German village Wolfsheim in 1235. The Erfurter chronographer claimed that 'it seems right that the blood of a person who thirsts for blood, should be spilled, according to this prophecy "since you hate blood, blood shall pursue you" (Ezekiel 35: 6).' See *Annales Erphordenses*, in MGH Scriptores, 16 (Hanover, 1859) p. 31; C. Ocker, 'Ritual Murder and the Subjectivity of Christ: a Choice in Medieval Christianity', *Harvard Theological Review*, **91** (1998) 183.
24. *Mahzor Vitry*, ed. S. Hurwitz (Jerusalem, 1963 [reprint]) p. 296.
25. Habermann, *Sefer Gezerot*, pp. 201–2.
26. The Hebrew is: 'lakhen tokhalkhem esh haholkhim mi vami', referring to Exodus 10: 8.
27. According to the Rome manuscript of *Sefer Nizahon Yashan*, the quotation from Ezekiel is accurate: you ate man, and not 'the' man. This manuscript also omits the accusation concerning the eating of Christian children.
28. Berger, *The Jewish–Christian Debate*, no. 244.
29. Joseph ben Nathan Official, *Sepher Joseph Hamekane*, ed. J. Rosenthal (Jerusalem, 1979) no. 36.
30. Yom Tov Lipmann Mühlhausen, *Sefer Hanizahon* (Altdorf-Nürnebrg, 1644; reprinted: Jerusalem, 1984) no. 119.
31. Sigmund Freud suggested that the Christian communion contains traces of the original sin, in which the 'expelled brothers' murdered their Ur-father-God (replaced by his Son) and 'ate' him (*Totem and Taboo* [London, 1950] pp. 154–5).
32. C. W. Bynum, *Holy Feast and Holy Fast* (London, 1987) p. 133.
33. See Genesis 22: 10 (Hebrew: *ma'achelet*); Exodus 3: 2.
34. BT Sanhedrin 99a: 'There shall be no Messiah for Israel, because they 'ate' him in the days of Hezekiah'.

35. For example, the *piyut* 'Elohim Zedim' on the persecutions of 1096: 'they ate us, and they destroyed us' (Hebrew: *ahalunu wa-yehalunu*). See Habermann, *Sefer Gezerot*, p. 84.

36. S. Grayzel, *The Church and the Jews in the XIIIth Century*, vol. 1, revised edn (New York, 1966) no. 18, pp. 115–17; S. Simonsohn, *The Apostolic See and the Jews: Documents, 492–1404*, Pontifical Institute of Medieval Studies: Studies and Texts, 94 (Toronto, 1988) no. 82, pp. 86–8. The same accusation was repeated in 1346 in a synod of the archdiocese of Prague (S. W. Baron, *A Social and Religious History of the Jews*, vol. 9 (Philadelphia, PA, 1965) p. 26). On the regulation to receive communion once a year, at Easter, see Miri Rubin, *Corpus Christi: The Eucharist in Late Medieval Culture* (Cambridge, 1991) pp. 64–73.

37. Baron, *Social and Religious History*, p. 252.

38. Mishna Avoda Zara 2: 1; BT Yevamot 114a; Tosefta Shabbath 9: 22.

39. *Or Zarua*, II, Piske Avoda Zara, no. 146. See also S. Lieberman, *Tosefta Ki-fshutah*, Part III: Mo'ed (Shabbath) (New York, 1962) pp. 149–50; M. M. Kasher, *Torah Shelemah*, 9 (Exodus) (New York, 1954) pp. 246–7. The employment of Christian wet-nurses was less common among Jews in Spain than in Ashkenaz, and this is well reflected in the Halachic attitudes. Whereas the Sephardic rabbis adopted the Talmudic opinions that prohibited hiring non-Jewish wet-nurses, Ashkenazic rabbis permitted it. Accordingly, only in Ashkenaz (and France) did the rabbis have to require that the Christian wet-nurses ate kosher food. I thank Rami Reiner for sharing his knowledge on this issue with me.

40. Earlier this same year Innocent III raised the charge that a Christian scholar had been found murdered in a Jewish latrine (Grayzel, *The Church*, no. 14, pp. 108–9). The Talmudic passage regarding the status of Jesus in the next world is quoted by Nicholas Donin (Ch. Merchavya, *The Church versus Talmudic and Midrashic Literature, 500–1248* [Hebrew] (Jerusalem, 1971) pp. 276–7). In the year 1096, a Jew in Mainz, David ben Netanel, cursed Jesus in the presence of a Christian crowd by quoting this same passage (Neubauer and Stern, *Hebräische Berichte*, p. 11). See also Chr. Cluse, 'Stories of Breaking and Taking the Cross: a Possible Context for the Oxford Incident of 1268', *Revue d'Histoire Ecclésiastique*, 15 (1995) 418.

41. I. G. Marcus, *Rituals of Childhood: Jewish Acculturation in Medieval Europe* (New Haven, CT, 1996) pp. 74–127. The quotation is on p. 108.

42. I. Ta-Shema, 'The Earliest Literary Sources for the Bar-Mitzva Ritual and Festivity' [Hebrew], *Tarbiz*, 68 (1999) 586–98.

43. Ibid., p. 590; H. Schirmann, 'Kinot on the Persecutions in the Land of Israel, Africa, Spain, Germany, and France' [Hebrew], *Kobez Al Jad*, 3(13) (1939) 58–62.

44. Habermann, *Sefer Gezerot*, pp. 198–201.

45. When questioning the motivation behind the blood libels in the sixteenth century, Samuel Usque writes: 'O nations, why do you not consider how forbidden and abominable it is for a Jew to eat blood? . . . How, then, can you charge the Jews with killing a child to take his blood? It can only be an affliction which the Lord permits the Jews; misfortunes which the Lord wishes to give them; persecutions which their sins bring on . . . payment in kind for the commandment of blood which they and their fathers wilfully

broke' (*Consolation for the Tribulations of Israel*, trans. and ed. by M. A. Cohen (Philadelphia, PA, 1964) p. 213). He surely means the failure of the *marranos* to keep the commandment of circumcision. Here again we find the principle that blood stands for blood.

46. Y. Even Shmuel, *Midreshei Ge'ulah* (Jerusalem and Tel Aviv, 1954) p. 282.
47. I. J. Yuval, 'Jews and Christians in the Middle Ages: Shared Myths, Common Language', in R. S. Wistrich (ed.), *Demonizing the Other: Antisemitism, Racism and Xenophobia* (Singapore, 1999) pp. 88–107.
48. D. Malkiel, 'Infanticide in Passover Iconography', *Journal of the Warburg and Courtland Institutes*, 56 (1993) 85–99.
49. Successful struggles by Jews against blood libels are often discussed in the book; however, not a single case of host desecration accusation is mentioned. See *Shevet Yehudah*, ed. Baer and Shohat, ch. 7, pp. 26–9, 32–3; ch. 8, pp. 46–50; ch. 12, pp. 56–7; ch. 16, pp. 62–3; ch. 17, pp. 63–6; ch. 29, pp. 72–4; ch. 38, pp. 92–4; ch. 61, p. 126. An intensification of Jewish efforts to refute the blood libels around 1500 is also evident in Abrabanel (see n. 22 above), Samuel Usque (n. 45 above) and Elijah del-Medico, *Sefer Behinat Hadat*, ed. J. J. Ross (Tel Aviv, 1984) pp. 98–9.
50. On this discussion see F. I. Baer, 'New Notes on Shebet-Jehuda' [Hebrew], *Tarbiz*,6 (1935) 152–79; A. A. Neuman, 'The Shevet Yehudah and Sixteenth-Century Jewish Historiography', *Louis Ginzberg Jubilee Volume*, the American Academy for Jewish Research (New York, 1945), English section, pp. 253–73; D. E. Carpenter (ed.), *Alfonso X and the Jews: an Edition and Commentary of Siete Partidas 7–24 'De los Judios'* (Berkeley and Los Angeles, CA, 1986) pp. 64–5.
51. A similar episode took place in Neustadt (near Nürnberg) in 1421. See I. J. Yuval, 'Juden Hussiten und Deutsche nach einer hebräischen Chronik', in A. Haverkamp and F.-J. Ziwes (eds), *Juden in der Christlichen Umwelt während des späten Mittelalters* (Berlin, 1992) p. 98. On the non-violent image of the Jews in modern historiography see E. Horowitz, '"The Vengeance of the Jews was Stronger than their Avarice": Modern Historians and the Persian Conquest of Jerusalem in 614', *Jewish Social Studies*, 4/2 (1998) 1–39.
52. These arguments are repeated in various cases throughout the book, and they are also attributed to a Muslim ruler who was summoned for a meeting with the king (ch. 17, p. 64).
53. Mishna Baba Quama 4: 3.
54. Republished by M. Stern, *Andreas Osianders Schrift über die Blutbeschuldigung* (Kiel, 1893; repr. Berlin 1903).
55. Po-Chia Hsia, *The Myth of Ritual Murder*, pp. 124–31, 136–43.

6
Anti-Jewish Violence and the Place of the Jews in Christendom and in Islam: a Paradigm

Mark R. Cohen

Introduction

In an earlier publication,[1] I attempted to explain why Islamic–Jewish and Christian–Jewish relations followed such different courses in the Middle Ages. More specifically, I endeavoured to clarify, in greater detail and more effectively than had been done in the past, why Jews lived so much more securely in the Arab–Islamic world than in Christendom. In the context of this conference on 'Religious Persecution with Reference to Christians and Jews' I would like to invert the comparison in order to explain why anti-Jewish violence was more severe in the Christian orbit.

In the book, I concentrated on the northern Latin lands, the locus of the most severe anti-Jewish persecution in medieval Christendom and the region providing the starkest contrast with the world of Islam. Indeed, so great did that dissimilarity seem to nineteenth-century Jewish historians that they exaggerated the medieval Muslim–Jewish relationship, especially in Spain, into a 'Golden Age' or 'myth of an interfaith utopia'. More recently this 'myth' has been challenged in some circles by what I call a 'countermyth of Islamic persecution' or a 'neo-lachrymose conception of Jewish–Arab history'. This revisionism transfers the 'lachrymose conception of Jewish history' in medieval Christendom (to use the phrase made famous by the late Jewish histor-ian Salo W. Baron) to the medieval Arab world. The 'neo-lachrymose conception' arose in apologetic response to Arab exploitation of the myth of the interfaith utopia as a weapon against modern Zionism. It claims that Muslim–Jewish relations in the Middle Ages were much

107

worse than previously thought, at times even as bad as the gloomy plight of the persecuted Jews of Christendom.[2]

Some reviewers of my book, as well as some colleagues who did not review it, questioned my choice of northern, Latin Europe for the comparison with Islam. A few even took me to task for comparing the 'best' period of Islamic–Jewish relations with the 'worst' case of Christian–Jewish relations, or asked whether I was not comparing apples with oranges.[3] It appears that the brief justification I offered for leaving out such areas as Italy, Reconquista Spain, Byzantium and medieval Poland was not adequate. In addition, the fact that I distinguished between the *early* Middle Ages, when Jews enjoyed relative security in northern Europe, and the period beginning with the crusades (eleventh century), when conditions deteriorated, was apparently insufficient to demonstrate my awareness of differences *within* medieval Latin Christendom. Largely overlooked, too, perhaps because of the admittedly small space I devoted to them, were my digressions on southern France (the Midi), where the forces operating in society were different from those in the north, and so was the level of anti-Jewish violence.[4]

This conference, and particularly the questions about 'other Christendoms' it evoked in discussion, encouraged me to address these issues in this paper. For heuristic reasons, I continue to stand by the choice I made. The most fruitful way to illuminate differences in a comparative study, it still seems to me, is to take cases that are palpably different. A remote, if not bizarre, analogy would be the case of wine. In order to fathom why, say, expensive French wine pleases while cheap American wine does not, it would make sense to select a French wine that people generally believe to be excellent, and an American wine that people generally deem poor in quality, and submit each to chemical tests to identify the ingredients that makes them so different. This approach would be especially appropriate if, say, 'wine revisionists' claimed that even the best French wine was not as good as people think, and that in some ways it was as bad as bad American wine, if not worse.

But there is another value to the method, and this has occurred to me only in retrospect, or rather, the idea was only latent when I wrote the book. Looking back, I believe the argument of the book constructs a paradigm, or 'ideal type', that can be applied to other periods and regions of Christendom (and even to later medieval Islam). With that in mind, after describing the paradigm below, along the lines of the book, I will make some comments about 'other Christendoms' with the goal of showing that levels of anti-Jewish violence were lower precisely where one or more of the forces that, according to the paradigm, contributed to the extreme violence in northern, Latin Europe in the high and later Middle Ages were absent or at least attenuated.

Anti-Jewish violence in Christendom and Islam

It is hardly necessary to prove that Jews were victimized by persecution and violence in the medieval Christian world, particularly in northern Latin Europe, the so-called Ashkenazic lands, and especially beginning with the First Crusade. Well known are the many instances of large-scale massacre of individuals or groups of Jews; of the torture or execution of Jews on charges of killing and exsanguinating Christian children, of poisoning wells, or stealing and 'torturing' the Eucharist wafer (the 'host desecration libel'). In another kind of violence, Jews were exposed to measures intended to weaken the hold of Judaism and convert them to the majority faith. These acts included official limitation of Jewish occupational opportunities, the burning of the Talmud, compulsory attendance at conversionary sermons, and other violations of the age-old recognition of the right to practise Judaism without interference. Finally, there were the expulsions – from kingdoms, counties and towns – beginning in the twelfth century, expulsions that left most of Latin Europe '*judenrein*' by the year 1500.

Persecution and violence were not absent from the world of Islam. But they were much less violent and occurred much less frequently than the victimization of the Jews in Christian lands. Most episodes of oppression consisted in efforts to enforce the restrictive laws of the Pact of 'Umar, a document ascribed to the second caliph but more reasonably representing the cumulative practice of the first century or so following the death of the Prophet Muḥammad.

In return for adherence to certain demeaning regulations plus payment of an annual poll tax, Islam guaranteed non-Muslims protection for their persons, property and buildings, as well as freedom of religion. Monotheists who had received a divinely revealed scripture were awarded this status and were designated 'protected people', *ahl al-dhimma*, or *dhimmīs*. Out of pragmatism, dualist Zoroastrians in Persia, whose scripture, the Avesta, contained teachings of the prophet of ancient Iran rather than revelations by one of their gods, and polytheist Hindus in India, were also assimilated to the *dhimmī* class.

Instances of Islamic violence against Jews begin with the expulsion from Medina of two Jewish tribes that rejected Islam, followed by the massacre of the male members of a third tribe, the Banū Qurayẓa, which was accused of treachery. But most of the recorded incidents of Islamic oppression in post-Muhammadan times did not target Jews specifically, but rather the broader, multiconfessional class of *dhimmīs*. The persecution of *dhimmīs* by the Egyptian caliph al-Ḥākim during the first two decades of the eleventh century represents one of the most infamous episodes of violence against Christians and Jews. The caliph's aggression included such violent acts as the destruction of churches and synagogues and compelling the *dhimmīs* to choose between conversion to Islam (which is forbidden by Islamic law) and expulsion. Al-Ḥākim was thought by medieval Arab historians to be mad. They recognized that some of his outrageous impositions exceeded the canonical stipulations of the Pact of 'Umar.[5] At the end of the reign of terror, al-Ḥākim and his son and successor permitted Jews and Christians to return to their original faiths. Since this, too, contravened Islamic law, as a face-saving measure the regime made the converted *dhimmīs* pay arrears in the poll tax.

The Almohad persecutions in North Africa and Spain that began in the middle of the twelfth century similarly went beyond the original intentions of Islamic law by forcing Jews *and* Christians to adopt Islam or flee from the realm. Thousands who elected to stay and hold fast to their religion were killed.

Nonetheless, the Almohad persecutions demonstrate that violence against Jews per se (as opposed to Jews as members of the *dhimmī* class) was exceedingly rare in medieval Islam. The well-known persecution and forced conversion of Jews in Yemen in the latter part of the twelfth century was in practice, if not in theory, anti-Jewish, for Jews were the only *dhimmīs* living in Yemen at the time.

The most discussed instance of Islamic violence aimed at Jews per se was the 'pogrom' that struck the Jews of the Spanish–Muslim Berber kingdom of Granada in 1066. Joseph ibn Nagrela, the Jewish vizier who succeeded his illustrious father, Samuel the Nagid, was assassinated by his enemies at court on 31 December 1066. The story is related in the 'memoirs' of Sultan Abdallah – grandson of Joseph's patron, Sultan Bādīs – in a graphic, eyewitness description of the background to the events and of the massacre itself.[6]

The pogrom of 1066 may appear to the modern reader to be a case of medieval 'antisemitism', especially because it seems to have been incited by a rabidly anti-Jewish Arabic poem urging the murder of Joseph and the Jews of Granada. But as Bernard Lewis has pointed out correctly, this particular type of anti-Jewish violence derived, not from the irrational hatred of the Jews that we associate with antisemitism, even medieval Christian antisemitism,[7] but from the contractual nature of the *dhimma*, as codified in the Pact of 'Umar. Even the nasty Arabic poem attacked the Jews because 'they have violated our covenant with them'.[8] Sultan Abdallah's complaint about 'the notorious changes which [the Jews] had brought in the order of things, and the positions they occupied, in violation of their pact' expresses the notion of a hierarchical society that could not tolerate any alteration in the 'right order' of society, especially when perpetrated by infidels.[9] The Jewish historian Abraham ibn Da'ud, writing his *Book of Tradition* in Spain a century later, mentions the pogrom of 1066 briefly and criticizes the behaviour of Joseph ibn Nagrela, saying that he was 'haughty – to his destruction'.[10]

Constructing the paradigm

Employing a comparative perspective, we can construct a paradigm that explains why anti-Jewish violence was more severe in Christendom than in Islam. We shall see that while religious hatred, channelled against the Jews by certain elements of the clergy, constituted a major cause of anti-Jewish violence in Christendom, other forces, especially economic, legal and social ones, could either exacerbate or moderate the innate religious intolerance of both societies in which the Jews lived. We shall see that exacerbating forces dominated in Christendom while moderating ones prevailed in Islam. However, as we shall also demonstrate, where moderating forces were present in some domains of Christendom, there, too, anti-Jewish violence was at a lower level.

Religious attitudes towards the Jewish minority[11]

Religious attitudes stood at the foundation of hostility towards the Jews in both Christianity and Islam. But theological factors, as well as historical context, contributed to anti-Jewish violence much more profoundly in Christendom, particularly northern Christendom, than in Islam. Why?

Christianity originated as a rebellious sect within Judaism, committed to the belief that Jesus was the Messiah spoken of in the Hebrew Bible, and for some, the son of God. Jesus was crucified, and his crucifixion, though carried out by Roman soldiers, was blamed by Christians on the Jews. Despite the positive redemptive interpretation assigned by Christ's disciples to his death in the concept of the vicarious atonement, the act of killing Jesus continued to be considered unforgivable, to be remembered in perpetuity.

In the early centuries, while Rome denied Christians the legal recognition long conferred upon the Jews, and periodically persecuted devotees of the suspect new sect, Christianity grew apart from Judaism. Christians, beginning with Paul, depicted Jewish law as worthless, at least for Gentile converts.[12] Subsequently, Christianity replaced Jewish law with a new nomism, the nomism of Christ. In this scheme, Mosaic law appeared as but a transient stage on the way to the final, spiritual, messianic perfection in the Incarnation. Christians also constructed the notion that they formed a New Israel, replacing *Old* Israel, which, rejected by God, now lay defeated, frozen in time. This doctrine aimed at answering Jewish and Roman charges that Christianity represented a revolutionary innovation. In defence of their beliefs, Christian thinkers asserted that they were not innovations at all. Everything had been anticipated in the Old Testament, which needed simply to be understood in a spiritual or allegorical sense. This argument could theoretically appeal to Romans, who, in the Greek tradition, admired allegory. The Church's claim to represent the New (and True) Israel, fulfilling God's word in the Old through a New Testament, introduced a bold credo that eventually undermined Judaism's place in the divine scheme of history.

At the end of the third century, Emperor Diocletian cracked down on the Christians with a ferocity that spawned a new breed of martyr. Some Church writings accused the Jews of wreaking their own violence against the Christians. Exaggerated or not, an indelible memory of Jewish persecution of Christians became embedded in Christian consciousness. This memory, consistent with paradigmatic Old Testament accounts of Israelite animosity towards Moses and the prophets, and also with New Testament stories about Jewish persecution of Jesus and his disciples, would much later nourish irrational fantasies about Jewish violence towards Christians, invoked to justify a militant Christian response, both verbal and physical, to Judaism and the Jews.

The opportunity to take revenge upon the Jews arrived when Emperor Constantine converted to Christianity in the early fourth century and then made it the official religion of the Roman Empire. Roman imperial laws from the time of Constantine onwards weakened the tolerationist legislation of the pagan period with anti-Jewish rulings and prejudicial statements about Judaism.

Empowered after centuries of marginality, some Christians asked why the Jews, rejected by God, deprived of their temple and country, and left with a worthless set of laws, should continue to exist among Christians. What purpose did they serve, if any, in history?

Augustine (354–430) answered these questions by articulating a positive role for the Jews in Christian salvation history. This was his doctrine of 'witness', which served over the following centuries to justify the preservation of the Jews within Christendom. God wanted the Jews to be preserved because, in their defeated state, the Jews bore witness to the triumph of Christianity. In addition, by preserving the Old Testament, the Jews bore witness to the authenticity of its Hebrew text, the sacred repository of prophecies of Christ's life and death. The Hebrew original, alongside the Jewish Greek translation, the Septuagint, provided a ready answer to the Gentiles when they accused the Christians of making up these prophecies. The final act of witness, already enunciated by Paul, would ensue when, at the second coming of Christ, the Jews converted to Christianity.

Whatever the aetiology of Augustine's doctrine,[13] it became adopted as papal policy for protecting the practice of Judaism and helped safeguard the Jewish community from physical destruction during the early Middle Ages, up to the eleventh century. Thereafter, anti-Jewish violence erupted with regularity. Before examining this change, it is useful to compare Christianity with Islam, where similar foundations for anti-Jewish violence were *not* laid and no theological rationalization had to be created to protect the Jews from physical extirpation.

First, the founder of Islam claimed neither messiahship nor divinity. While the Jews of Medina ridiculed him in his lifetime, Muḥammad died a natural death. Thus, unlike Christians, Muslims had no grounds for holding the Jews culpable for the demise of their progenitor. True, biographical accounts of the Prophet and Muslim traditions (*ḥadīths*) depict Jewish attempts on the Prophet's life.[14] These stories surfaced whenever Muslims looked for reasons to mistrust contemporary Jews. Without a 'propheticide', however, and lacking an iconographic tradition like Christianity's that might have provided the illiterate Muslim masses with a graphic representation of Jewish enmity towards Muḥammad in Medina, the Islamic–Jewish conflict could not generate the kind of tension and hatred that so inflamed the conflict between Christianity and the Jews.

Furthermore, early Islam, in contrast to early Christianity, did not have to struggle for centuries to gain recognition from a hostile and powerful enemy like Rome. After Medina and further conquests in the south, Islam carried the day as the established religion of Arabia, and the confidence instilled by this victory propelled the new religion to greater triumphs as it went on relatively quickly to overcome the two huge but weakened empires of Byzantium and Persia. One result was that Islam never needed to portray itself as a 'New Israel'. Nor did it need a doctrine of witness to justify the continued existence of Judaism or Christianity in its midst and its protection of these communities from violence. In addition, Islam did not stake a claim to the scriptures of the Jews and

Christians. It did not need these books to prove its own claims to the truth. Shared claim to scripture laid the foundation for continual tension between Christianity and Judaism over the interpretation of the message of Jewish holy writ and promoted centuries of interreligious polemics seeking to weaken Judaism's grip on its adherents. Islam, by contrast, did not claim possession of either the Hebrew Bible or the New Testament. In fact, it dismissed the existing texts of these scriptures as corruptions of their original, divinely inspired teaching, which was restored in the Qur'ān. The foretellings of Muḥammad in the Old and New Testament that form one of the themes of Islamic polemical literature are but a pale imitation of the much more indispensable Christian method of Old Testament exegesis. In its interreligious polemics, Islam was if anything more hostile towards Christianity than towards Judaism because Christianity, with the apparent polytheism of its Trinity, was more repugnant to strictly monotheist Islam. In brief, the religious character of Islam and the historical circumstances of its origins with respect to Judaism attenuated violent anti-Jewish feelings and illuminate, by contrast, the religious causes of Christian violence against the Jews in the Middle Ages.

The legal position of the Jews[15]

A comparison of the legal position of the Jewish minority in Christendom and in Islam teaches much about the differing degrees of anti-Jewish violence in these two societies. In the Christian world, the Jews were affected by different, often conflicting, laws – the canon law of the Church, feudal custom, the law of the developing medieval state, and the law of the emergent city. The mixture of laws to which the Jews were subject sometimes worked to their advantage, as when they benefited from the more favourable ordinances of one institution or another. But, for the most part, the legal situation of the Jews expressed itself in a certain arbitrariness and irksome unpredictability. Moreover, the Jews had a unique legal status: they were the *only* infidels living within northern Christian society, especially after the last pagans in eastern Europe were converted. As Jews became more vulnerable to Christian violence beginning with the crusades, secular rulers tightened their jurisdiction. A *jus singulare* developed for the Jews. They became monarchical 'property'–'serfs of the royal chamber' as they came to be called in Latin – subject to a special and oppressively restrictive legal status. The unmediated legal relationship between monarch and Jew continued to provide some measure of badly needed protection in an increasingly hostile environment. But it further underscored their alien status. It gave rulers licence to exploit them through heavy taxation and extra-ordinary exactions and permitted them to place limits on the Jews' free-dom of movement. The 'enserfment' of the Jews to secular rulers was complemented by an old, patristic doctrine concerning the 'perpetual servitude' of the Jews. Revived by Pope Innocent III in 1205 to compete with secular claims of control over the Jews, it somewhat weakened the otherwise reliable protection that canon law and papal decrees had afforded the Jews since the beginning of the Middle Ages and added justification to their legal subservience. The legal dependence of the Jews and their ultimate isolation from the law that encompassed the majority Christian society ultimately led to their removal from Chris-tendom during the widespread expulsions of the thirteenth to fifteenth centuries.

In the Islamic world the Jews were not subject to a unique legal status. They were not the only infidels on the scene. There were plenty of Christians, many more than there were Jews, and Persia contained a significant Zoroastrian population. All of them *dhimmīs*, they were sub-sumed under the same legal umbrella, subject to (not isolated from) the same unified law that governed Muslims, the *sharī'a*, or Islamic Holy Law, enjoying protection by the Islamic state in return for the annual poll tax payment and adherence to restrictions that suited their lowly religious position *vis-à-vis* Islam. With the exception of the poll tax, however, the restrictive laws were often circumvented by the *dhimmīs* with the tacit approval of the authorities, at least before the general decline in Jewish status that began around the twelfth to the thirteenth centuries.

Residing as it did in a unitary corpus, the *sharī'a*, *dhimmī* law was essentially consistent, predictable, and not given to arbitrary interpret-ation and application. The relative stability over time of the basic law regarding the treatment of non-Muslims thus assured the Jews a consid-erable degree of security against violence.

The economic factor[16]

If the legal situation of the Jews weakened their defences against Chris-tian anti-Judaism, Jewish status was also adversely affected by Christian economic views and economic development in northern Europe. Ideo-logically, early Christianity emphatically disapproved of the accumula-tion of wealth, especially profit (*turpe lucrum*, 'shameful gain') acquired through commerce.[17] During the barbarian period, the merchant's very way of life gave offence to those living the stable, sedentary existence of the dominant rural economy.[18] As Jews at that time predominated in international commerce and as most Jews were travelling merchants,[19] Christian theological disdain of the Jew was complemented by the general suspicion of the Jew-as-merchant. Adhering to a mistrusted profession, the Jew epitomized the quintessential 'stranger', as described in a famous essay by sociologist Georg Simmel.[20]

The stigma attached to the suspect, roving, stranger-merchant did not deter Christian rulers from encouraging Jewish long-distance traders from settling permanently in their domain. Motivated by the principle of utility, the Carolingian kings set out to induce itinerant Jewish merchants to settle permanently in their realm by issuing them a version of the standard Carolingian charters of privileges. In addition to generous dispensations to facilitate their commercial activities, these patents guaranteed Jews the all-important right to live by their own laws.[21] This conformed to the pluralistic premise underlying the principle of the personality of law that the Franks, more than other Germanic groups, perpetuated in the early Middle Ages. As in the pagan Roman world, the pluralism of the tribal world of the Germanic barbarians spelled advantages for the Jews. The relatively secure position of the Jews in northern Europe during the early Middle Ages owes much to this feature of the social order.

The revival of urban life in the late tenth and eleventh centuries was to have a damaging impact on the Jews, even as a more favourable view of commerce came into vogue during the twelfth century to accommodate the needs of the new Christian urban commercial class. Gradually, Christian traders in northern Europe squeezed their Jewish competitors out of the market economy by excluding them from the developing commercial guilds.[22] Already accustomed to putting out some of the surplus capital generated by their commercial transactions in the form of loans, Jews were now compelled to transfer their main energies to moneylending. And other Jews who for various reasons were pushed out of productive occupations also found their only means of livelihood in making loans to Christians.

Christian debt to Jewish moneylenders had a decidedly detrimental affect on Jewish–Christian relations in northern Europe. Though important for the growth of the economy and for the relief of distress, and grudgingly tolerated for those reasons, moneylending contributed greatly to anti-Jewish feelings and even to violence. From the end of the twelfth century, Jew-hatred intensified as a result of the association of usury in Christian minds with the inimical twin evils of heresy and the devil. Christian moneylenders abounded, but Christian borrowers detested Jewish creditors with special vigour, for they could not easily abide their economic dependence on creditors whom they regarded as infidels and Christ-killers. Not surprisingly, around this time Jewish usury began to occupy an important place in the growing number of disputations and in Jewish apologetical and exegetical works.[23]

If in Christendom economic attitudes and economic development worked against the Jews, in Islam they helped safeguard them from persecution and violence. To begin with, Islam appeared on the historical scene at a time when commercial exchange over considerable distances was still entrenched in the conquered areas of southwest Asia and North Africa. Moreover, pre-Islamic Arabs had themselves participated in the caravan trade linking spice-rich southern Arabia with the markets of Egypt, Byzantium and Syria. Thus it is not surprising that early Islam expressed a positive attitude towards commerce. The ideological encouragement of trade, firmly established in both the Qur'ān and the *ḥadīth*, carried over into the realm of Islamic law as it gradually took shape in the hands of merchant-jurists during the early Islamic period.

With an urban middle class appearing in well-established Islamic cities centuries before similar developments in northern Europe, the long-distance trader, including the Jewish merchant, could hardly be viewed as an alien. And the commercial duties Jews paid, though discriminatory, were at least predictable and usually assessed fairly, in stunning contrast to the arbitrary 'fiscal terrorism' (to use William C. Jordan's coinage, referring to France and England) that Jews in northern Europe suffered.[24]

For the most part, and certainly until the late Middle Ages, the Jews of Islam did not fulfil economic functions that accentuated the lowly status assigned them by Islamic religion and law. Our main source, the documents of the Cairo Geniza, probably give disproportionate weight to merchants, because it is they, more than others, who wrote letters filled with, or soliciting, information that was needed to conduct business affairs with a minimum of uncertainty or risk. But the Geniza also proves that Jewish economic life was widely diversified. Predominantly urban dwellers, the Jews of Islam were well integrated into the economic life of society at large and without occupational stigma.

Unlike in Europe, where Jewish fortunes ran in inverse relationship to the general economic well-being of society – up when commercial economy was down (in the early Middle Ages); down when Christendom was economically on the rise (in the high Middle Ages) – in Islam, Jewish economic rise and decline, alike, *coincided* with economic trends in the general society. This was one sign among many of their embeddedness in the economic and social order of their Muslim surroundings.

Moreover, in a revealing contrast with Europe, the main arena for Jewish credit transactions in the Islamic world during the early and high Middle Ages was *within* the Jewish community – meeting the need for capital as in all advanced economies.[25] There was, to be sure, some lending on interest to the non-Jew, which biblical law expressly permitted, even though it was discouraged by the Geonim, the heads of the *yeshivot* (religious academies) in Islamic Babylonia, as it was by most of their rabbinical counterparts in Europe.[26] But the reverse was also true: Jews regularly borrowed at interest from Muslim moneylenders. In short, the equation, 'usury begets anti-Jewish hatred and anti-Jewish violence', did not apply in the Muslim world as it did in Christendom.

The place of the Jews in the social order[27]

Another perspective on anti-Jewish violence in medieval Christendom and Islam makes use of anthropology to explore the relationship between Jews and the majority society. In *Homo Hierarchicus*,[28] the French social anthropologist Louis Dumont, studying the caste system in India, draws some important implications for the social order of premodern societies. Dumont argues that the fundamental idea unifying societies composed of a multiplicity of groups and statuses is *hierarchy*; that hierarchy, more than power, determines how the elements of such a society interact. Hierarchy is 'the principle by which the elements of a whole are ranked in relation to the whole, it being understood that in the majority of societies it is religion which provides the view of the whole, and that the ranking will thus be religious in nature'.[29] The hierarchical relationship is that 'between encompassing and encompassed or between ensemble and element'.[30] Paradoxically, the elements of caste systems – and, by extension, the components of any stratified society – manage to coexist more or less harmoniously precisely because each knows that it and all the other subgroups of the population are part of a totality. Differences are accepted as natural in a fully formed, or 'ideal-type' hierarchical society, of which the Indian caste system is the best-known example.

To hierarchy must be added the element of marginality. As refined by certain sociologists,[31] 'marginality theory' describes a type of hierarchy in which members of a group '(1) do not ordinarily qualify for admission into another group with which, over varying lengths of time, it is more or less closely associated; (2) ... differ significantly in the nature of their cultural or racial heritage; and (3) [have between them] limited cultural interchange or social interaction'.[32] In a 'marginal situation' – unlike caste systems with their ideal of permanent or total group exclusiveness – there is some permeability of the 'barriers' (or boundaries) separating elements in the hierarchy.[33] While both marginality and exclusion engender or reflect intergroup tensions, marginality expresses a somewhat less alienated relationship between the subordinate group and the larger society.[34] Medieval Christians had ways of talking about the social order and the place, if any, of the infidel or the non-conforming Christian in that order. They were conscious of *status* rather than *class*. Thinkers spoke in terms of hierarchies – who was above and who was below.[35] In medieval Christendom, the Jews can be said to have started out at the bottom of the hierarchy, but in a marginal situation. As the historian Bernhard Blumenkranz has shown,[36] there was considerable social and economic interchange between Jews and Christians during the centuries between the barbarian invasions and the rise of the crusading spirit in Latin Europe. Spatially, the Jews were not excluded; they lived relatively close to and peacefully with their Christian neighbours. They pursued a variety of occupations, not just long-distance commerce, and, here and there, Jews even held public office.

With the rise of the crusading spirit and the deepening of Christian consciousness and piety in the population at large beginning in the eleventh century, Jews gradually began to lose the benefits of their marginal situation and came slowly but decisively to be excluded from the hierarchy of the Christian social order. By the thirteenth century, Jews in the Latin west no longer conformed to Dumont's model of hierarchy based on relations 'between encompassing and encompassed or between ensemble and element'. Increasingly, Christians felt that Jews threatened to enfeeble Christian society. The universalism of the encompassing whole, with its place, however lowly, for the Jews, had by that time, as Jacques Le Goff observes, been tempered by a 'Christian particularism, the primitive solidarity of the group and the policy of apartheid with regard to outside groups'. None of the complex models of subdividing Christendom into socio-professional 'estates', which increasingly came to characterize the social order from the beginning of the thirteenth century, had any place for the Jews.[37] Exclusion was, so to speak, the 'final solution' for the Jews in medieval Catholicism, and it was carried out in one of three violent ways: forced conversion, massacre and, most effectively, the expulsion of most of western European Jewry from Christian lands by the end of the fifteenth century.

Turning now to medieval Islam, the question can be asked: What of hierarchy and marginality there? It is possible to read the Pact of ʿUmar as a document imposing exclusion on the *dhimmīs*, since it requires that they distinguish themselves from Muslims by special garb and by certain other behaviour. In reality, however, the regulations of the Pact were intended not so much to exclude as to reinforce the *hierarchical* distinction between Muslims and non-Muslims within a single, encompassing social order.[38] Non-Muslims were to remain 'in their place', avoiding any act, particularly any religious act, that might challenge the superior rank of Muslims or of Islam. The *dhimmī*, however, occupied a definite rank in Islamic society – a low rank, but a rank nevertheless. Marginal though they were, the Jewish (and Christian) *dhimmīs* occupied a recognized, fixed and safeguarded niche within the hierarchy of the Islamic social order. In Bernard Lewis's words, they held a kind of 'citizenship', though as second-class citizens, to be sure.[39]

Actual social relations, shaped by the marginal situation of the *dhimmīs*, somewhat alleviated the discriminatory content of the regime of humiliation imposed in the Pact of 'Umar. Here, research on the structure of Islamic society offers insight regarding the Jews. On the basis of sources for tenth- and eleventh-century Iraq, Roy Mottahedeh describes how clerks, soldiers and merchants (as well as physicians) formed recognizable groups manifesting mutual loyalties that bound them to one another in the pursuit of common interests.[40] These 'categories', as Mottahedeh calls them, were not guilds of the European variety with strict confessional criteria for admission. Indeed, the Muslim world had nothing like the guilds of the medieval west.

Dhimmīs, like Muslims, could be found in nearly all categories of Islamic society, working alongside Muslims who outranked them by virtue of their religion. In these situations, the regime of differentiation in the *dhimma* laws let everyone know who was a non-Muslim and who a Muslim. The 'loyalties of category', to use Mottahedeh's terminology, extending across the Muslim–non-Muslim distinction, doubtless softened the discrimination taught by Islamic religion.

Additional explanations for the relatively more favourable position of the Jewish minority in medieval Islam compared to their brethren in medieval northern Christendom emerge when viewing the Jewish–Muslim relationship through the lens of ethnicity. Historically, ethnic heterogeneity has been much more characteristic of the medieval Orient than the medieval Occident. Arabs, Iranians, Turks, Kurds, Berbers, Jews, Christians, Zoroastrians and others populated the social landscape, composing a 'mosaic' that gave society a richly hued human and cultural texture. Further, as noted already, the *dhimmī* group exhibited heterogeneity within its own ranks, with two (in some places three) nonconforming religions coexisting in the same space.[41]

Clifford Geertz, in his study of economic life in Sefrou, Morocco, lists as one of the basic givens or 'characteristic ideas' of the 'mosaic' pattern of social organization in Middle Eastern society the fact 'that non-Muslim groups are not outside Muslim society but have a scripturally allocated place within it'. Diversity thus has a preservative function.[42] A. L. Udovitch and Lucette Valensi have illustrated this culture of differences in traditional Islamic society through field-work among a modern vestige of classical Jewish-Arab interethnic coexistence on the island of Jerba, Tunisia.[43]

Complementing the model of hierarchy and marginality, then, these anthropological and sociological findings help explain what in the medieval Middle East appears to be a 'tolerant' relationship between Muslims and non-Muslims. By contrast, it explains the absence of tolerance and the growth of anti-Jewish violence in medieval Christendom. As of the twelfth century, Europe experienced an exclusivism growing from religious and proto-national homogeneity in medieval Catholicism. This aggravated existing anti-Jewish feeling and begot a mounting level of anti-Jewish violence. Christendom in northern Europe from this period on lacked the ethnic differentiation which in Islam worked, along with religious, legal and economic factors, to preserve the Jews' niche in the hierarchy of the social order and to nurture the social, economic and cultural embeddedness of the Jewish minority in Arab society. These factors kept the Jews from being totally excluded from the Islamic social order, mitigated the perception of them as aliens, and safeguarded them from the type and severity of violence that plagued Jews especially in the northern Christian lands for the better part of the high and later Middle Ages.

Memory of persecution[44]

Not surprisingly, and in stark contrast with their brethren in Christian lands, who constructed their history as a long chain of suffering, the Jews of the Islamic Middle Ages preserved very little collective memory of Muslim acts of violence, hardly anything that smacks of a 'lachrymose conception of history'. For example, whereas the story of the massacre of the Banū Qurayẓa and the expulsion of the other Jewish tribes of Medina is related in detail in the Arabic biography of the Prophet and in Islamic chronicles, the event is recalled only once by a medieval Jew, in a brief and oblique allusion in Maimonides' consolatory 'Letter on Forced Apostasy' (*Iggeret ha-shemad*), written around 1165. It took the modern Hebrew poet Saul Tchernikowsy to make the ordeal famous in his poem called 'The Last of the Banū Qurayẓa'. The Almohad persecution comes in for brief mention in Ibn Da'ud's *Book of Tradition*, and underlies three other literary works: a treatise by Maimonides and another by his father, consoling contemporary Jews who had converted to Islam under duress, and in a chapter of an ethical work written somewhat later by Joesph b. Judah ibn ʿAqnin. The author of the last-mentioned, who had converted to Islam, blames the apostasy itself for the continued oppression.

Jews in Ashkenazic lands composed myriads of poems, elegies and chronicles in the wake of persecution and martyrdom, many of which entered the liturgy and are still recited in synagogues today. By contrast, among the thousands of Hebrew poems written during the classical Islamic centuries, I know of only one example of a poetical reaction to an outbreak of Islamic persecution: Abraham ibn Ezra's eulogy for the Jewish communities in Spain and North Africa annihilated by the Almohads in 1147–8. Ibn Ezra had spent several years travelling around the Jewish communities of Ashkenaz and quite possibly was influenced by the Ashkenazic model. The only other examples of Hebrew elegies about persecution written by Andalusian poets refer to acts of violence perpetrated by *Christians*, not Muslims. Two elegies on the death of Joseph ibn Nagrela in 1066 by a contemporary Hebrew poet lack the faintest allusion to the gruesome, violent circumstances surrounding his demise. The al-Ḥākim episode, with its killings, destruction of synagogues and forced conversions, is mentioned in one contemporary letter found in the Cairo Geniza but it has no literary parallel to the narrative chronicles of persecution penned by Jews in Christian lands, or even to the records of Egyptian Christians, who underwent the same ordeal. On the other hand, at the height of that persecution, when a Muslim mob attacked a Jewish funeral procession and 23 Jews were imprisoned, one of the captives, a noted liturgical poet, wrote a short Hebrew treatise, modelled on the biblical Megillah of Esther and celebrating the Jews' release from prison by order of the same unpredictable caliph.

Only in the later Middle Ages did the Jews of Islam begin to record episodes of maltreatment, during a period when their situation in the Muslim world had deteriorated. This setback parallelled the general decline in economy and society following the rise of European commercial power in the Mediterranean and of the Mongols in the east, and the growing influence of hard-line Muslim clerics who objected to the continued influence of *dhimmīs* in the civil service. Still, these episodes were far less violent than anti-Jewish brutality in Christian lands. Moreover, even as their humiliation increased, Jews, with little exception, continued to be allowed to live among Muslims without compulsory conversion or threat of expulsion, let alone mass murder.

Does this mean that the Jews of Islam did not feel oppressed, even persecuted at times, like their brethren in Christian lands? Far from it. But, I believe, they did not experience it as an unbreakable chain of persecution. They experienced Muslim contempt, along with their Arab–Christian neighbours. But they had substantial confidence in the *dhimma* system. If they kept a low profile, if they paid their annual poll tax, then they expected to be protected – not to be forcibly converted to Islam, not to be massacred, and not to be expelled. When the system broke down, as it did under al-Ḥākim, under the Almohads, and under the Shiite regime in Yemen in the twelfth century, Jews felt the impact of the violence no less than the Ashkenazic Jews of Europe. But they recognized these as temporary failures of the *dhimma* arrangement, just as some Arabic chroniclers of al-Ḥākim's reign wrote disparagingly of his excesses. Moreover, as Goitein has noted, Jews in twelfth-century North Africa and Muslim Spain probably recalled how al-Ḥākim and his son had formally permitted the forcibly converted to return to their former faith, and expected the same of the Almohads, correctly, as it turned out.[45] Doubtless this helps explain why, under threat, Jews in Islamic lands favoured superficial conversion (like the Islamic *taqiyya* recommended for Muslims faced with persecution) over martyrdom, unlike their self-immolating Ashkenazic brethren, who had little hope of being officially allowed to return to Judaism after their baptism.

Applying the paradigm to 'other Christendoms'

The paradigm that emerges from the comparative study of Christian–Jewish and Muslim–Jewish relations in the Middle Ages is complex. It claims that anti-Jewish violence is related, in the first instance, to the totalitarianism of religious exclusivity. Historically, totalitarianism of religious exclusivity characterized both Islam and Christianity. But anti-Jewish violence was more pronounced in Christendom because innate religious antagonism was combined with other erosive forces. The first of these lay in economic circumstances that excluded the Jews from the most respected walks of life. The second lay in legal status, namely, the evolution of a special law for the Jews and a system of baronial or monarchical possessory rights that could be manipulated in an arbitrary manner. Economic marginalization and a special, arbitrary legal status, combined with another adverse factor, social exclusion, to rob the Jews of their rank in the hierarchical social order. The gradual replacement of the ethnic pluralism of Germanic society of the early Middle Ages by a medieval type of nationalism, parallelling the spread of Catholic religious exclusivity to the masses, also contributed to the enhancement of the Jew's 'otherness' and to his eventual exclusion from western Christendom. This exclusion was accomplished by violence, in the form of murder, forced conversion, or, most successfully, expulsion.

The paradigm proves its usefulness when applied to the Islamic world, where diminished anti-Jewish violence correlated with a lower level of religious intolerance, a less arbitrary legal status under the protection of religious law, the absence of monarchic possessory rights, widespread Jewish economic differentiation, greater social inclusiveness, and ethnic and even religious pluralism. But if the value of the paradigm is truly to stand up to scrutiny, even Christendom should show lower levels of anti-Jewish violence where some or all of the factors discussed above in connection with northern Latin Europe were altered – where they bore greater similarity to the economic, legal and social circumstances of the Jew in medieval Islamic society. Below, we offer, briefly, some suggestive observations along these lines.

Northern Europe in the early Middle Ages

As the paradigm predicts, Jews in northern Europe in the early Middle Ages experienced relatively little anti-Jewish violence. Although identified largely with international trade, Jews displayed a certain amount of economic differentiation, as Bernhard Blumenkranz has argued. Secondly, the legal status accorded the Jews strongly emphasized monarchical protection, rather than, as developed later on, monarchical possession. Royal jurisdiction over the Jews manifested, not arbitrariness, but consistent favouritism – directly related to the utility Jewish long-distance merchants offered to secular princes and their courts. Furthermore, law in the ethnically pluralistic Germanic period adhered to the principle of personality (rather than territoriality), and Jews represented just another ethnic group with its own tribal law. Jews were socially more included in northern Europe in the early Middle Ages than they were later on. Indeed, adumbrating Church objections to such favourable Jewish status that would help erode it later on, Agobard, the archbishop of Lyons, railed against the liberties of social intercourse encouraged by the privileges granted Jewish merchants by the Carolingian rulers in the ninth century.[46]

Religious totalitarianism was not yet established in northern Europe during the early Middle Ages. Much of society still clung to its pre-Christian, Germanic (pagan) tribal religious ways. Massacres of the Jews and other types of persecution reared their ugly head only in the eleventh century, and especially beginning with the First Crusade. As the paradigm suggests, this shift of attitude and treatment of the Jews corresponded with several important changes: the penetration of Catholic exclusivity to the lower classes; the rise of Christian commerce, forcing Jews out of the commercial marketplace and into despised moneylending; the growth of constraining possessory rights; and the decline of ethnic pluralism.

Southern France and Italy

Turning our attention now to southern France (the Midi), and also to Italy, the paradigm again proves its heuristic value. Mediterranean Latin Christendom, most agree, offered a much more hospitable environment to Jews than the northern reaches of Europe.[47] There is no hard evidence for persecution of the Jews in the south of France during the First Crusade,[48] and Gavin Langmuir has pointed to the absence of the ritual murder libel in southern France, particularly Languedoc.[49] In general, Jewish communities of the south lived in more placid, integrated fashion in their surroundings than their northern European brethren. They were less segregated from Christians, and their economic activities varied, from moneylending to small- and large-scale trade, including long-distance commerce. Jews also worked as toll-gatherers in association with Christians, and in land-transfer brokerage.[50] In Languedoc Jews also engaged in many other occupations, such as agriculture (either as landowners or as tenant farmers), artisanry, commerce (butchery, cereals), peddling, brokerage, medicine and public offices. And they owned immovable property, whether in the form of agricultural land, houses, or artisan or commercial establishments.[51] Though moneylending was the dominant profession, it seems to have had less dire consequences for Christian–Jewish relations in the Midi than in the north.

If heterogeneity in Jewish economic life was a crucial agent tempering anti-Jewish feeling, so were other features peculiar to southern Europe, and here, Italy may also be included. In southern Europe, continuity with the Roman past and a sharper memory of Roman legal traditions and perseverance of Roman legal procedures contributed to the relative security of the Jews as compared with their status in the northern communities, where Roman law was virtually forgotten in the early Middle Ages. Similarly, the antiquity of Jewish settlement in the south – bordering on indigenous habitation and resembling the native status of Jewry in Arab lands – contributed to a more tolerant atmosphere.

To this should be added the general attitude towards urban life. In the north, urban autonomy, revived after centuries of decay, ran counter to feudal preferences. In the Mediterranean lands, by way of contrast, urban society of the Roman era had never quite died out. Moreover, unlike its counterpart in northern countries, the nobility in southern

Europe was not cut off from city life.[52] Continuity of urban life, absence
of rigid social boundaries between city and countryside, and an aristoc-
racy receptive to town habits correlate with a greater openness towards
the Jew in the south. As in the Islamic world, urbanism fitted more
organically into the social order of Mediterranean Christendom than in
the north, where the town represented a considerable disruption in the
traditional pattern of social life and organization.

Rooted in the south since Roman antiquity, Jews comprised a more
organic part of the urban landscape, as they did in the Islamic world. In
the cities of Languedoc during the twelfth to the fourteenth centuries,
for example, as in the Islamic world, residential segregation was minimal. Jewish communities in the Midi reaped more of the benefits and
experienced fewer of the liabilities of corporate status than did the Jewish communities of England and royal France.[53] With William Jordan,
we may also point to the absence of regional political unification in the
south, which elsewhere was accompanied by intensified degradation of
the Jew-as-alien. As the paradigm predicts, all these contrasts with the
situation of the Jews in northern Europe served to temper anti-Jewish
violence in the Midi. It was only after the conquest of southern France
by the French monarchy that some of the anti-Jewish oppression characteristic of northern European Christendom began to appear in these
annexed lands.

Reconquista Spain

The omission of medieval Christian Spain from previous discussion does
not mean that Spanish Christendom belies the paradigm. Quite the contrary. Particularly during the period of the Reconquista, when conditions
for the Jews were deteriorating in the northern European heartland,
Jewish–Christian relations in Spain were relatively tolerable. In accordance with the paradigm, Jews displayed considerable economic differentiation. Yitzhak Baer's gleanings from Spanish archival documents
in *Die Juden im christlichen Spanien*,[54] which underlay his *History of the
Jews in Christian Spain*, confirm that Jews worked in a wide variety of
occupations – , in commerce, agriculture, handicrafts, medicine, as well
as in service to the Spanish courts.[55] The legal status of the Jews, though
similar in principle to the 'Jewish serfdom' in France, Germany and
England, was less injurious in practice. While this improved juridical
situation certainly owed something to the example of the less oppres-
sive *dhimma* system in the regions of Muslim Spain (Andalusia) annexed
by the Catholic conquerors, it also had much to do with the utility Jews
provided the kingdoms of Aragon and Castile in the administration and

taxation of the colonized Muslim territories. In the *fueros*, or town charters specifying immunities or exemptions granted by the king or lord, Jews were accorded a large measure of equality with Christians and Muslims, especially during the early period of the Reconquista. Christian Spain, too, exhibited pluralism in the mixture of Catholics, Jews and Muslims that composed its society and influenced its culture, appropriately labelled the culture of 'convivencia'.

Things began to deteriorate with the conclusion of the Reconquista and the concomitant decline of Jewish utility, as Baer argues. Beginning with the pogroms of 1391, Jews in Christian Spain had to weather anti-Jewish violence akin to that experienced by their brethren in northern Latin lands. Mass conversions to Catholicism produced the well-known Marrano problem and led at the end of the century to the establishment of the fierce Spanish Inquisition, which, though charged with prosecuting Catholic heresy in the form of Marrano Judaizing, indirectly undermined the security of unconverted Jews as well. As a result of the marriage between Ferdinand, king of Aragon, and Isabella, queen of Castile, a policy of religious unification was pursued in the Iberian peninsula with the expulsion of the Jews in 1492, the conquest of Granada and sub-jugation of its Muslims in the same year, and then the suppression of Islam in 1502 (Castile) and 1525 (Aragon).

Byzantium

Like the Jews in Mediterranean France, on the one hand, and in Christian Spain during the Reconquest, on the other, Jews in the Byzantine domains (which for several centuries following Justinian's successful imperial expansion in the mid-sixth century included southern Italy and Sicily), experienced less violence in the Middle Ages than in the Latin west, especially during the period before the Fourth Crusade in 1204.[56] And, not surprisingly, the Jews' economic, legal and social position conformed to characteristics expected according to the paradigm.

Byzantium was the locus of early Christian–Roman Jewry law – in the Theodosian and Justinianic Codes. Despite the theological disparagement of Judaism and attendant social animosity that crept in, these corpora clung tenaciously to the tolerationist features inherited from pagan Roman legislation. This benefited the Jews even more than in southern Europe, for Roman law had a continuous life in the late antique eastern Empire and its medieval Byzantine successor.[57] Moreover, the Latin Christian model of 'Jewish serfdom', with its monarchical possessory rights over the Jews and attendant arbitrariness, did not make signifi-cant inroads into the eastern Roman Empire. In addition, the evidence portrays a Jewish population with a differentiated economic profile, even in the later period.[58]

Ethnic and religious pluralism characterized the Byzantine Empire, in some places until its very end in the fifteenth century. Armenians, Catholic Christians, Jews (both Rabbinate and Karaite) and in some parts of the Empire also Muslims (e.g. Anatolia) occupied the same social space in the Byzantine domains, and this diffused the natural hostility towards the 'other', to the advantage of the Jews.

As in southern France, urban centres were never eradicated in the Byzantine Mediterranean. They presented relatively comfortable places for Jews to inhabit, and for the same reasons. Moreover, Jews had lived in Byzantium from pre-Christian Roman times and constituted a more embedded element in society than the tiny, alien, immigrant communities from both shores of the Mediterranean that forged the virgin Jewish settlements in northern Europe during the early Middle Ages.

Medieval Poland

Medieval Poland also exhibits the applicability of the paradigm. In the thirteenth century, Polish kings invited German townsmen from the west to settle in their land in order to revive urban and commercial life there. Favourable legal conditions were offered, to Christians in the form of the liberal law code of the city of Magdeburg, to Jews in the form of protective royal charters. The least restrictive version of the German charters for Jews was chosen as the model, and the language of 'Jewish serfdom' was omitted, as Polish kings side-stepped the harsh policies against the Jews then insinuating themselves in the west under the pressure of the above-mentioned papal doctrine of 'perpetual servitude of the Jews'.[59] In Poland, Jews found expansive economic opportunities during the period of initial settlement that liberated them from exclusive reliance on moneylending and its untoward consequences in Christian animosity. Economic diversification, reaching a high degree with the stepped-up immigration of western Jews in the fifteenth and sixteenth centuries to the adjacent and gradually merging kingdom of Poland and duchy of Lithuania (formally united in 1569), made the Jews less 'other' and further helped attenuate anti-Jewish violence.[60] The Polish Commonwealth, especially in its geographically expanded form, represented a large, multi-ethnic kingdom of Lithuanians, Poles, Armenians, Ukrainians (Orthodox Christians in distinction to the Catholic Poles), Tatars and Jews. Pluralism, as the paradigm asserts, constituted an advantage for the Jews, as it did elsewhere in the Middle Ages.[61]

The paradigm applied to late medieval Islam

The paradigm applies not only to 'other Christendoms' but also to Islam in the late Middle Ages. In this period, the thirteenth to the fifteenth centuries, and in some places beginning as early as the twelfth, forces that had moderated Islamic intolerance in the earlier period weakened and Jews experienced greater oppression. In Morocco, Jews emerged from the catastrophe of the twelfth-century Almohad persecutions as the *only* non-Muslim religious minority group (Christian converts to Islam did not revert to their former religion). Thus, the pluralism of the earlier period, when Jews, Christians and Muslims occupied the same physical space and anti-Jewish hostility was diffused among the two non-Muslim groups, ended there. Economically, a general decline set in in most of the Islamic world in these centuries, and this, too, had an adverse effect on Jewish well-being. With the growth of an Islamic form of 'feudalism', better, statism, in the Mamluk Empire of Egypt and Syria-Palestine (1250–1517), tighter economic controls meant less economic freedom for the merchant and artisan classes, and, as a minority group, the Jews necessarily fared worse than Muslims.

The principal factors eroding the security of the Jews in the thirteenth to fifteenth centuries affected the Christians as well, indeed even more seriously. Political developments were decisive here: the invasion and occupation of the Levant by the crusaders beginning at the end of the eleventh century, and the Mongol conquests in the thirteenth century in the eastern Islamic world. In both cases, non-Muslims (especially Christians in the case of the crusaders) were suspected of collusion with the enemy, or at least of tacit support, and this raised Muslim anxieties, with ensuing harsh treatment. It should be noted that fear of non-Muslim treachery had some rational basis, unlike the situation in the Latin west.[62] Helpless in their European dispersion and lacking loyalty to an external enemy state, the Jews became *imagined* enemies of Christendom, allies of the devil, and inveterate, recidivist Christ-killers, allegedly murdering Christian children and also desecrating the host, poisoning wells, and wreaking other atrocities against Christians and Christianity.

Pursuing the paradigm further, the Islamic world experienced an economic upswing in the sixteenth century, owing to the Ottoman conquests in the Levant, Egypt and North Africa. This more or less coincided with the expulsion of the Jews from Christian Spain. Many thousands of the Iberian exiles resettled in the Arab and Turkish lands of the Muslim world, where they were welcomed by the Ottoman rulers because of their commercial skills and international contacts and because of their ingrained enmity towards the Ottomans' own foreign Christian foes. They thus contributed significantly to the renewed florescence of a monetary, commercial economy on the eastern and southern shores of the Mediterranean. This buoyed up a hitherto languishing Jewish economy.

Even though certain factors changed in the late Islamic Middle Ages – and the paradigm helps explain how this impacted adversely on Jewish life – things did not reach the low point they did in northern Christendom or Reconquista Spain at the end of the fifteenth century. Pluralism, which never died out in the Islamic world the way it did in Latin Christian lands, assured a certain amount of protection to the Jews in the Ottoman period and even earlier, especially in places where Christians and Jews continued to coexist. The most characteristic causes of Christian anti-Jewish violence in the west did not occur in the Islamic world, even in this later period. Irrational antisemitism expressed in Christendom in the blood libel and the host desecration accusation were absent in Islamic lands. The Ottoman cases of ritual murder accusation (a handful in the early modern period and a proliferation in the period of Ottoman stagnation in the nineteenth century) were almost without exception incited by Christians, and the Ottoman government steadfastly rejected them.[63] Jews did not come to be identified with the devil, nor with heresy, for neither the devil nor heresy had the same salience in Islam as in Christianity. Conversions to Islam, especially by Christians, increased in the late Middle Ages, and often the neophytes were suspected of opportunism and subjected to a kind of 'inquisition'.[64] But professing Christians and Jews continued to hold positions of power in Muslim governments, even during the period of decline, and despite the vigorous complaints of Muslim clerics.

The humiliating provisions of the Pact of 'Umar intensified in application as the well-being of the Muslim masses declined in the late Middle Ages. But pressures here and there to cancel the protection granted non-Muslims by the *dhimma* system to the contrary notwithstanding, the Pact of 'Umar stood fast within the Islamic *sharī'a* to safeguard Jews and Christians from the kinds of violent excesses that struck Jews in Christian lands. Mass murder did not plague the minority communities in Islam as it did the Jews in Latin Europe. Occasional zeal to destroy non-Muslim houses of worship was often held in check by the application of due process of Islamic law.[65] With little exception, Islam did not employ the strategy of expulsion to rid itself of Jews and other religious minorities.

Life was more difficult for Jews and Christians in the later Islamic Middle Ages, to be sure, and the paradigm helps explain this. But Jews continued to hold their place in the social order of Islamic society. It was a lowly rank, a marginal position, to be sure, and it was accompanied by considerable humiliation. But it was nonetheless a recognized rank. Unlike Christendom, which solved its Jewish problem in the later Middle Ages by murder, forced conversion or expulsion, none of these violent 'solutions' to the 'Jewish problem' came to be employed in the Islamic world, for Islam continued to accept the Jews as an embedded and organic element of society, even as the general climate of well-being and security of the earlier period waned.

Notes

1. *Under Crescent and Cross: The Jews in the Middle Ages* (Princeton, NJ, 1994).
2. I discuss the historiographical issue in the first chapter of *Under Crescent and Cross*, which expands upon two earlier essays, 'Islam and the Jews: Myth, Counter-Myth, History', *The Jerusalem Quarterly*, **38** (1986) 125–37, and 'The Neo-Lachrymose Conception of Jewish–Arab History', *Tikkun*, **6** (1991) 55–60. The first of these articles was reprinted more than once and also appeared twice in Hebrew, in B. L. Sherwin and M. Carasick (eds), *The Solomon Goldman Lectures*, vol. 5 (Chicago, 1990) pp. 20–32; in S. Deshen and W. P. Zenner (eds), *Jews and Muslims: Communities in the Precolonial Middle East* (Basingstoke, 1996) pp. 50–63; translated into Hebrew in *Zmanim: A Historical Quarterly* (Tel-Aviv University), **9** (1990) 52–61; and in a revised Hebrew version in: H. Lazarus-Yafeh (ed.), *Muslim Writers on Jews and Judaism: The Jews among their Muslim Neighbours* (Jerusalem, 1996) pp. 21–36.
3. Even before the book appeared, this critique surfaced. See, for example, N. Stillman's rejoinder, 'Myth, Countermyth, Distortion', to my article in *Tikkun* (previous note), and also the exchange in the 'Letters to the Editor', *Tikkun* (July 1991). See also Steven Bowman's review of the book in *American Jewish Studies Review*, **21** (1996) 397.
4. Robert Chazan's critique in *Medieval Stereotypes and Modern Antisemitism* (Berkeley and Los Angeles, CA, 1997), pp. 142–3, n. 5, misrepresents the contrast I drew between northern Europe, where Jews arrived (as I write) as roving, alien, long-distance merchants, and hence found an ambivalent reception in those parts, and southern Europe, where their more comfortable existence coordinates, among other things, with near indigenous habitation, resembling the 'native status' of the Jews of Arab lands (*Under Crescent and Cross*, p. 103). The findings of the multi-factored approach of my work, and, in particular its comparison of northern Europe with the world of Islam, summarized in this paper, yields, I believe, more nuanced understanding of the comparative fate of medieval Jews in Christendom and Islam than Chazan's suggestion that differences 'simply reflect distinctions between well-settled and newcomer communities' (Chazan, p. 143).

5. *Under Crescent and Cross*, p. 74.
6. English translation of the long passage in B. Lewis, *Islam: From the Prophet Muhammad to the Capture of Constantinople*, vol. 1 (New York, 1974) pp. 123–34. See also *Under Crescent and Cross*, p. 165.
7. See G. I. Langmuir, *History, Religion and Antisemitism* (Berkeley and Los Angeles, CA, 1990), ch. 13 ('Religious Irrationality') and ch. 14 ('From Anti-Judaism to Antisemitism').
8. B. Lewis, 'An Anti-Jewish Ode: the Qasida of Abu Ishaq against Joseph Ibn Nagrella', in S. Lieberman in association with A. Hyman (eds), *Salo Wittmayer Baron Jubilee Volume* (Jerusalem, 1975) pp. 657–68 (English section); M. Perlmann, 'Eleventh-Century Andalusian Authors on the Jews of Granada', *Proceedings of the American Academy of Jewish Research*, **18** (1948–9) 269–90; N. Stillman, *The Jews of Arab Lands* (Philadelphia, PA, 1979), pp. 214–16, reprinting the translation by Bernard Lewis. See *Under Crescent and Cross*, pp. 165–6.
9. This concept is developed in *Under Crescent and Cross*, ch. 6, and summarized below.
10. Abraham ibn Da'ud, *Sefer ha-Qabbalah*, ed. and trans. G. D. Cohen (Philadelphia, PA, 1967) pp. 75–6 (English translation).
11. This section summarizes the argument of Chapter 2 in *Under Crescent and Cross*.
12. See J. Gager, *The Origins of Anti-Semitism: Attitudes toward Judaism in Pagan and Christian Antiquity* (New York, 1983) pp. 193ff. Regardless of Paul's true convictions in this matter, what is important for our present purpose is that subsequent Christian thinkers understood him to have preached that the coming of Christ had abrogated the Law, and this became salient in the Jewish–Christian conflict.
13. See, recently, P. Fredriksen, 'Divine Justice and Human Freedom: Augustine on Jews and Judaism, 392–398', in J. Cohen (ed.), *From Witness to Witchcraft: Jews and Judaism in Medieval Christian Thought* (Wiesbaden, 1996) pp. 29–54 and Cohen's own *Living Letters of the Law: Ideas of the Jew in Medieval Christianity* (Berkeley and Los Angeles, CA, 1999), pp. 19–65.
14. Stillman, *The Jews of Arab Lands*, pp. 129–30. As background to the expulsion of the Jewish tribe of the Banūu Naḍīr from Medina, a story is told in al-Wāqidī's *Kitāb al-maghāzī* (Book of the Military Campaigns [of the Prophet]) about some Jews from that tribe who tried, unsuccessfully, to kill Muḥammad by dropping a stone upon him from the roof of a house, in hopes, thereby, of putting an end to his religious movement. Although such attempts failed, it is nonetheless noteworthy, by way of contrast with the Christian–Jewish religious conflict, that murderous intentions towards the founder of Islam did not play a particularly important role in the Islamic polemic against the Jews in the Middle Ages.
15. For the extended argument underlying the summary that follows see *Under Crescent and Cross*, chs 3 and 4.
16. The summary below is based on Chapter 5 in *Under Crescent and Cross*.
17. J. W. Baldwin, *The Medieval Theories of the Just Price: Romanists, Canonists, and Theologians in the Twelfth and Thirteenth Centuries*, transactions of the American Philosophical Society (Philadelphia, PA, 1959) pp. 12, 13 (monograph reprinted in *The Evolution of Capitalism. Pre-Capitalist Economic Thought: Three Modern Interpretations* (New York, 1972).

18. See, for instance, on Christian disdain for the merchant, N. Zacour, *An Intro-duction to Medieval Institutions* (Toronto, 1969) pp. 54–5, and H. Pirenne, *Medieval Cities: Their Origins and the Revival of Trade*, trans. F. D. Halsey (Garden City, NY, 1956) pp. 86–7.

19. Latin sources are specific on this. Jews are frequently referred to by such phrases as 'mercatores, id est Judei et ceteri mercatores' ('merchants, that is, Jews and other merchants'); J. Aronius, *Regesten zur Geschichte der Juden im fränkischen und deutschen Reiche bis zum Jahre 1273* (reprint, New York, 1970), p. 52 (no. 122, dated *c.* 906); also p. 55 (no. 129): 'ne vel Judei vel ceteri ibi [in urbe Magadaburg] manentes negotiatores' ('nor Jews nor other mer-chants remaining there [in Magdeburg]'). The view of Bernhard Blumen-kranz that Jews were more broadly spread across the economy (*Juifs et Chrétiens dans le monde occidental 430–1096* [Paris and the Hague, 1960] pp. 12–33), a view meant to correct the hostile, antisemitic 'optical illusion' of Jewish predominance in monetary professions, is apologetically extravagant.

20. G. Simmel, 'The Stranger', in K. H. Wolff (ed. and trans.), *The Sociology of Georg Simmel* (Glencoe, IL, 1950) pp. 403–4.

21. These Latin charters, three of which are actually sample texts in a medieval collection of formularies (ed. K. Zeumer, in MGH, Formulae Merowingici et Karolini Aevi [Hanover, 1886] pp. 309–11, 325, along with the fourth char-ter (Latin text ed. M. Bouquet, in RHGF no. 232), are discussed in virtually every work dealing with the status of the Jews in the early Middle Ages. They are available in a new English translation in A. Linder, *The Jews in the Legal Sources of the Early Middle Ages* (Detroit, MI and Jerusalem, 1997) pp. 333–8, 341–3, 365–7. B. S. Bachrach's discussion of the charters in *Jews in Barbarian Europe* (Lawrence, KS, 1977), is also useful.

22. Wilhelm Roscher, a nineteenth-century historian whose theory about the Jews' role in medieval economic life still has merit today, wrote: 'In those days the Jews satisfied a great economic need, something which, for a long time, could not be done by anyone else, namely, the need for carrying on a professional trade. Mediaeval policy toward the Jews may be said to have followed a direction almost inverse to the general economic trend. As soon as peoples became mature enough to perform that function themselves, they try to emancipate themselves from such guardianship over their trade, often in bitter conflict. The persecutions of the Jews in the later Middle Ages are thus, to a great extent, a product of commercial jealousy. They are connected with the rise of a national merchant class.' English translation by Guido Kisch in his tribute to Roscher, 'The Jews' Function in the Evolution of Mediaeval Economic Life', in Kisch's *Forschungen zur Rechts-, Wirtschafts-und Sozialgeschichte der Juden*, vol. 2, p. 109 (originally published in *Historia Judaica*, 6 [1944] and later in Kisch's *The Jews in Medieval Germany* [1949; repr. New York, 1970] pp. 316ff. [= *Historia Judaica*, p. 5]).

23. Kisch, *The Jews in Medieval Germany*, p. 45, and p. 192; J. Trachtenberg, *The Devil and the Jews: The Medieval Conception of the Jew and its Relation to Modern Antisemitism* (Cleveland, New York, and Philadelphia, 1961) pp. 188–95; J. Rosenthal, 'The Law of Usury Relating to Non-Jews' [Hebrew], *Talpioth*, 6 (1952) 134; S. Stein, 'A Disputation on Moneylending between Jews and Gentiles in Me'ir b. Simeon's Milhemeth Miswah (Narbonne, 13th Cent.)', *Journal of Jewish Studies*, 10 (1955) 45–61.

24. Jordan's characterization is found in his *French Monarchy and the Jews: From Philip Augustus to the Last Capetians* (Philadelphia, PA, 1989) p. 153.

25. S. D. Goitein, *A Mediterranean Society: The Jewish Communities of the Arab World as Portrayed in the Documents of the Cairo Geniza*, vol. 1 (Berkeley, CA, 1967) pp. 256–7. For a typical case, or cases, of intra-Jewish borrowing and lending see M. A. Friedman, 'Responsa of Abraham Maimonides on a Debtor's Travails', in J. Blau and S. C. Reif (eds), *Genizah Research after Ninety Years: The Case of Judaeo-Arabic – Papers Read at the Third Congress of the Society for Judaeo-Arabic Studies* (Cambridge, 1992) pp. 82–92.

26. G. Libson, 'Between Jewish Law and Muslim Law: the Preference Given Neighbors of Differing Religions in Acquiring Neighboring Property' [Hebrew], *Pe'amim*, **45** (1990) 85.

27. See Chapters 6–8 in *Under Crescent and Cross*.

28. Louis Dumont, *Homo Hierarchicus: The Caste System and its Implications*, complete revised English edn, trans. M. Sainsbury, L. Dumont and B. Galiati (Chicago, IL and London, 1980). The French original appeared in 1966. For a discussion of the controversial reception of the book, see Dumont's 'Preface to the Complete English Edition'.

29. Ibid., p. 66 (italics in the original).

30. Ibid., p. 243 [in the author's 'Postface: Toward a Theory of Hierarchy'].

31. In the summary that follows I draw upon H. F. Dickie-Clark, *The Marginal Situation: A Sociological Study of a Coloured Group* (London, 1966) and N. P. Gist and R. D. Wright, *Marginality and Identity: Anglo-Indians as a Racially-Mixed Minority in India* (Leiden, 1973).

32. Gist and Wright, *Marginality*, p. 21.

33. Dickie-Clark, *The Marginal Situation*, pp. 32–3.

34. 'Marginality', as used here, differs from the way in which it is employed in the work of Maurice Kriegel, who, in turn, is also influenced by the model of the caste system of India. Writing about the Jews of Mediterranean Europe at the end of the Middle Ages, Kriegel uses 'marginality' more or less synonymously with 'exclusion'. He views the various moves to segregate Christians from Jews as acts to avoid contamination by Jewish 'impurity', as in a caste system requiring strict avoidance of those considered 'untouchable'. M. Kriegel, *Les Juifs à la fin du moyen âge dans l'Europe méditerranéenne* (Paris, 1979), esp. ch. 2. This approach has the disadvantage of obscuring the search for a more nuanced understanding of majority–minority relations regarding the Jews, one that makes it possible to account for or partially explain the differences between Christian–Jewish and Muslim–Jewish relations in the Middle Ages and also to steer away from the polarity of 'marginality-integration' that seems to underlie the critique of Kriegel by Noël Coulet in 'Les Juifs en Provence au bas moyen-âge: Les limites d'une marginalité', in *Minorités et marginaux en France méridionale et dans la péninsule ibérique (viie–xviiie siècles)* (Paris, 1986) pp. 203–19.

35. At the beginning of his history of the hierarchical concept of the 'three orders' (those who pray, who fight and who labour) in medieval Christendom, Georges Duby quotes words used by Pope Gregory the Great to describe the 'necessary inequality' of the social order: 'Providence has established various degrees and distinct orders [*ordines*] so that, if the lesser show deference to the greater, and if the greater bestow love on the lesser, then

true concord and conjunction will arise out of diversity. Indeed, the community could not subsist at all if the total order of disparity did not preserve it. That creation cannot be governed in equality is taught us by the example of the heavenly hosts; there are angels and there are archangels, which are clearly not equals, differing from one another in power and order.' See G. Duby, *The Three Orders: Feudal Society Imagined*, trans. A. Goldhammer (Chicago, IL and London, 1980) pp. 3–4.

36. B. Blumenkranz, *Juifs et Chrétiens dans le monde occidental 430–1096* (Paris, 1960). In English, Blumenkranz condensed his book into a pair of essays appearing in *The Dark Ages: Jews in Christian Europe, 711–1096*, part of the series, C. Roth (ed.), *The World History of the Jewish People* (Ramat Gan, Israel, 1966) pp. 69–99 and 162–74. In the chapter 'The Roman Church and the Jews', Blumenkranz writes: 'The Jews were part of medieval society, and not merely a group on the fringes.' But, like Kriegel, he equates 'on the fringes' with the shift to 'exclusion' that began, he says, in the eleventh century (ibid., pp. 98–9).

37. J. Le Goff, *Medieval Civilization, 400–1500*, trans. J. Barrow (Oxford, 1988) pp. 152, 261–4.

38. A. Noth, 'Abgrenzungsprobleme zwischen Muslimen und Nicht-Muslimen: Die 'Bedingungen 'Umars (*aš-šurūṭ al-'umariyya*)' unter einem anderen Aspekt gelesen', *Jerusalem Studies in Arabic and Islam*, 9 (1987) 290–315.

39. B. Lewis, *The Jews of Islam* (Princeton, NJ, 1984), p. 62.

40. R. Mottahedeh, *Loyalty and Leadership in an Early Islamic Society* (Princeton, NJ, 1980), pp. 108–15.

41. Carlton Coon paved the way for understanding the significance of ethnicity in Arab civilization: 'In the old Middle Eastern culture . . . the ideal was to emphasize not the uniformity of the citizens of a country as a whole but a uniformity within each special segment, and the greatest possible contrast between the segments. The members of each ethnic unit feel the need to identify themselves by some configuration of symbols. If by virtue of their history they possess some racial peculiarity, this they will enhance by special haircuts and the like; in any case they will wear distinctive garments and behave in a distinctive fashion. Walking though the bazaar you have no trouble identifying everyone you meet, once you have learned the sets of symbols. These people want to be identified. If you know who they are, you will know what to expect of them and how to deal with them, and human relations will operate smoothly in a crowded space. . . . This exaggeration of symbolic devices greatly facilitates business and social intercourse in a segmented society. It saves people from embarrassing questions, from "breaks", from anger, and from violence. It is an essential part of the mechanism which makes the mosaic function.' C. S. Coon, *Caravan: The Story of the Middle East*, rev. edn (New York, 1958), pp. 153, 167.

42. 'Middle Eastern Society . . . does not cope with diversity by sealing it into castes, isolating it into tribes, or covering it with some common denominator concept of nationality. . . . It copes . . . by distinguishing with elaborate precision the contexts (marriage, diet, worship, education) within which men are separated by their dissimilitudes and work, friendship, politics, trade where, however warily and however conditionally, men are connected by their differences.' C. Geertz, 'Suq: the Bazaar Economy in Sefrou',

in C. Geertz, H. Geertz and L. Rosen, *Meaning and Order in Moroccan Society* (Cambridge, 1979), p. 141.

43. A. L. Udovitch and L. Valensi, *The Last Arab Jews: The Communities of Jerba, Tunisia* (Chur, Switzerland, 1984).

44. See Chapter 10 in *Under Crescent and Cross*.

45. *Mediterranean Society*, vol. 2, pp. 299–300.

46. *Agobardi Lugdunensis Archiepiscopi Epistolae*, ed. E. Dümmler, in MGH, Epistolae Karolini Aevi, vol. 5 (Berlin, 1899) pp. 182–5, 190ff. An abridged English translation of Agobard's letters was published by K. R. Stow in *Conservative Judaism*, **29**, no. 1 (1974), esp. 63, 65.

47. A dissenting view is proffered by Maurice Kriegel. In his 'Un trait de psychologie sociale dans les pays méditerranéens du Bas Moyen Age: le juif comme intouchable', *Annales: économies, sociétés, civilisations*, **31** [1976] 326–30, and more comprehensively in *Les Juifs à la fin du moyen âge dans l'Europe méditerranéenne* (Paris, 1979), Kriegel argues – against received wisdom – that the attitude towards the Jew in southern Europe was just as hostile as it was in the north. Noël Coulet has challenged this revisionist thesis in '"Juif intouchable" et interdits alimentaires', in *Exclus et systèmes d'exclusion dans la littérature et la civilisation médiévales* (Aix-en-Provence, 1978) pp. 207–21. See also Coulet, 'Les Juifs en Provence au bas moyen âge: Les limites d'une marginalité', pp. 203–19.

48. Norman Golb put forth a provocative theory to the contrary, based principally on the identification in a Geniza letter of a nearly effaced place-name as Monieux, a city in Provence. N. Golb, 'New Light on the Persecution of French Jews at the Time of the First Crusade', *Proceedings of the American Academy for Jewish Research*, **34** (1966) 1–63 (the document in question had been published earlier, but the place-name interpreted differently, by J. Mann, *Texts and Studies in Jewish History and Literature*, vol. 1 [Cincinnati, OH, 1931] pp. 31–3). Recently a pair of scholars has shown that the place-name was much more likely Muño, a city in northern Spain, and hence the anti-Jewish violence had nothing to do with the First Crusade. E. Engel, 'The Wandering of a Provencal Proselyte: a Puzzle of Three Geniza Fragments' [Hebrew], *Sefunot* 7 (22) (1998–9) 13–21 and Y. Yahalom, 'The Muño Letters: the Work of a Village Scribe from Northern Spain' [Hebrew], ibid., 23–31. Golb's argument about a connection with the First Crusade was, to begin with, weak.

49. 'L'absence d'accusation de meurte rituel à l'ouest du Rhône', in M.-H. Vicaire and B. Blumenkranz (eds), *Juifs et judaïsme de Languedoc xiie siècle-début xive siècle*, Cahiers de Fanjeaux, 12 (Toulouse, 1977) pp. 235–49.

50. Y. Dossat, 'Les Juifs à Toulouse: un demi-siècle d'histoire communautaire', in ibid., pp. 132–5.

51. G. Nahon, 'Condition fiscale et économique des juifs', in ibid., pp. 63–72, following Saige, Luce, Régné, among others.

52. S. Reynolds, *An Introduction to the History of English Medieval Towns* (Oxford, 1977) p. 87, writes about the contrast between southern Europe, where 'a good many people lived in towns although they worked in the country', and the English case, where the separation of town and country was more marked.

53. W. Pakter, *Medieval Canon Law and the Jews* (Ebelsbach, 1988) pp. 20–5. Among reasons historians have given for 'the relative tolerance of southern

society in respect to the Jews', as summarized by William C. Jordan, is the fact that 'the Jewish communities of the Midi were an organic part of the cities and towns of the south. Jews had lived here since the period of Roman domination, and they had never lost the semblance of protection under Roman law'. See Jordan, *The French Monarchy and the Jews*, pp. 110–11.

54. *Erster Teil: Urkunden und Regesten*, 2 vols (Berlin, 1929–36; repr. [Farnborough], 1970).

55. Y. Baer, *A History of the Jews in Christian Spain*, 2 vol. (Philadelphia, PA, 1961–6).

56. On the earlier period, during which Jews of the Byzantine Empire experienced only four episodes of outright state persecution, see J. Starr, *The Jews in the Byzantine Empire, 641–1204* (Athens, 1939) and A. Sharf, *Byzantine Jewry from Justinian to the Fourth Crusade* (New York, 1971); and on the later period, S. Bowman, *The Jews of Byzantium, 1204–1453* (Alabama University, 1985).

57. Bowman, *The Jews of Byzantium*, pp. 98–9.

58. Ibid., pp. 117–21.

59. B. D. Weinryb, *The Jews of Poland: A Social and Economic History of the Jewish Community in Poland from 1100 to 1800* (Philadelphia, PA, 1973) pp. 35–50.

60. S. W. Baron, *A Social and Religious History of the Jews*, 2nd edn, vol. 16 (Philadelphia, PA, 1976) ch. 70 ('Socioeconomic Restratification').

61. G. Hundert, 'An Advantage to Peculiarity? The Case of the Polish Commonwealth', *Association for Jewish Studies Review*, 6 (1981) 21–38.

62. Notwithstanding the extravagant and unconvincing theory of Allen and Helen Cutler in *The Jew as Ally of the Muslim: Medieval Roots of Anti-Semitism* (Notre Dame, IN, 1986).

63. Uriel Heyd, 'Ritual Murder Accusations in 15th- and 16th-Century Turkey' [Hebrew], *Sefunot*, 5 (1961) 135–50; J. Landau, 'Ritual Murder Accusations and Persecutions of Jews in 19th-Century Egypt' [Hebrew], ibid., pp. 417–60; J. Barnai, '"Blood Libels" in the Ottoman Empire of the Fifteenth to Nineteenth Centuries', in S. Almog (ed.) and N. H. Reisner (trans.), *Antisemitism Through the Ages* (Oxford, 1988), pp. 189–94.

64. E. Strauss (Ashtor), 'L'inquisition dans l'état mamlouk', *Rivista degli Studi Orientali*, 25 (1950) 11–26.

65. M. R. Cohen, 'Jews in the Mamluk Environment: the Crisis of 1442 (A Geniza Study)', *Bulletin of the School of Oriental and African Studies*, 47 (1984) 425–48.

7
At the Frontiers of Faith

Gavin I. Langmuir

The title of the organization that so generously sponsored this conference embraces two religions; it might be well, therefore, to make the perspective in what follows clear. I shall not be relying on the beliefs of either Judaism or Christianity as premises. Nor shall I be trusting the theological rationalizations that justified the use of violence as descriptions or explanations of the violence any more than I would accept the Aryan myth as an explanation of why the Nazis acted as they did. I shall simply be analysing types of violence between Christians and Jews in unreligious, empirical and, some might say, crass terms in an effort to indicate their variety, characteristics and phases. I should also warn that I was asked to range widely; and to comply, I will start with the most basic question about the problem. Why has violence been such a prominent and horrible feature of the relations between Jews and Christians? Although long obscured by wishful thinking, the major answer is obvious. It was not any startling differences in how they conducted their daily life that caused the hostility; it was the competition of their gods.

No religious stance has caused more violence than monotheism,[1] for adherents of the different monotheisms are, at bottom, almost inescapably opposed, however quietly and tolerantly – or even, like John Hick,[2] extremely ecumenically. Monotheists not only worship a single god but also believe that their god exists independently of themselves and is the only god. But the claim of Christian theologians that Christians and Jews worship the same one and only god is contradicted by a very obvious historical fact: Jews and Christians have in fact believed and found satisfaction in different one and only gods. For if we think empirically, it is obvious that Jews have never worshipped a triune god, one of whose persons appeared on earth and was crucified. Conversely, Christians have never worshipped the god who commanded the 613 *mitzvot* (precepts) of the written and oral Torah of Judaism and who has not yet sent the Messiah.

The differences in the divinities worshipped are empirically obvious, yet both sides refuse to recognize what I call polymonotheism, the fact that different peoples can and do have their own different one and only gods, which are equally and equally monopolistically valid for them.[3] Consequently, adherents of each monotheism have judged what might be regarded simply as a major ethnic difference as a cosmic failure with the gravest consequences for human salvation. It is not surprising then that there has been serious friction and frequently great violence between Jews and Christians.[4]

What is religious violence? Motives for violence are almost always mixed; almost no violence in the Middle Ages was uninfluenced by religious beliefs; and some religious ideas were involved whenever Christians thought of Jews. Most major medieval violence and all violence against Jews could therefore be described as religious. Since that description is far too broad to provide a clear focus for analysis, we need a more restrictive conception. I will therefore define religious violence as the exertion of physical force on human beings, whether by societies or individuals, that is primarily motivated, and explicitly justified, by the established beliefs of their religion. I should note that this definition excludes verbal attacks, and that I will therefore not be discussing the religious polemics between Christians and Jews or the forced attendance of Jews at Christian sermons.

What inspires religious violence is the anger or fear aroused when believers who recognize that disbelief exists on the frontiers of their faith are seriously upset by the recognition that their faith is not unchallenged. For awareness of those frontiers makes societies and individuals conscious of the limits, whether physical or mental, to the sway of their religious beliefs and to the omnipresence and omnipotence they ascribe to their god, a recognition that can incite anger or fear or both. Before religious violence can occur, however, those who are angered must also be able to muster sufficient physical force to attack the people they deem a threat to their faith, which is why medieval violence between Christians and Jews was overwhelmingly Christian.

There were various frontiers of faith in medieval Europe where disbelief could inspire violence. They were both physical and mental. There were the physical barriers, both social and geographic, between the religions or religious communities to which Jews, Christians and Muslims belonged,[5] and there were the mental obstacles to faith within individuals. Awareness of any of those frontiers might be disturbing.

For Christians, it could be the simple knowledge that everyone did not, or not yet, belong to their religion and believe in their god. It could be the disagreeable knowledge that people in other religions who knew about Christianity, such as Muslims, rejected it. It could be the more irritating knowledge that members of a religious minority such as the Jews, who lived within the Christians' political society and were quite familiar with Christianity, remained unimpressed by it. It could be the yet more unsettling knowledge that some members of their own religious society, who had grown up with its beliefs, questioned some of those beliefs, for example, the eleventh-century Christians, both educated and uneducated, who questioned whether Christ was physically present in the consecrated wafer of the Mass.[6] Or it could be the very unsettling feeling that they themselves had questions about some beliefs of their religion.

The frontiers of faith did not exist uniformly throughout Europe. The existence of the major physical frontiers depended, of course, as in Spain, on the presence of Jews or Muslims in significant numbers. But I want to emphasize a different kind of variation that tends to be obscured by the history-of-ideas approach which explains medieval violence against Jews primarily by Christian theological ideas that were present throughout Europe. At no time was Christianity uniform across medieval Europe. What the Christian rituals, iconography, and the words expressing Christian beliefs meant to people differed at different times and in different places. A striking and highly relevant example is the fact that although there was little general hostility and little violence against Jews from 600 to 1096,[7] that changed dramatically with the massacres of 1096. What I want to stress particularly, however, is that, at any given time, the meaning of Christianity differed in varying degrees in the different regions of Europe.[8]

There was a variety within Europe of what I would like to call religious micro-climates, regions whose religious attitudes and actions differed significantly from each other because of their different languages, histories and cultural attitudes and their different political, social and economic conditions. To see what I mean, one has only to think, for example, of Celtic Christianity in seventh-century England, of the Cathars and the Albigensian Crusade in thirteenth-century Languedoc, of the absence of heresy in England before Wycliffe, or of the Beguine movement and the Rhineland mystics of the fourteenth century. And one of the differences between those micro-climates was the marked difference across Europe in attitudes towards Jews and in violence against Jews. In 1096, the first great massacre of Jews in Europe, like the violence accompanying the later accusations of ritual murder by cruci-fixion and ritual cannibalism, and of attacking Christ anew by torturing the consecrated host of the Mass, occurred almost exclusively in northern Europe.

If religious violence – according to my restrictive definition – first appeared in 1096,[9] violence on political frontiers that had an obvious religious component had been a familiar experience in Europe long before the eleventh century. Although secular motives were too involved in that violence before 1096 for it to fit my definition, it is undeniable that Christians frequently went to war against people of different religions. It has often been said that Christianity was basically against violence up to the eleventh century because clerics were forbidden to fight and Christian soldiers were canonically required to do penance for every person they killed, even in a just war, but that is misleading. Empirically, Christianity is what all the people who believed in some sense that Jesus was divine actually believed and did, as well as what ecclesiastics told them they ought to believe and do; and the difference between practice and precept was often large.

By 400, St Ambrose was attributing Roman victories to his god's help, and it is hard to discover wars that were declared unjust. Moreover, it is questionable how many Christian soldiers or warriors actually did penance between 400 and 1100. As Peter Partner has nicely put it in his recent study of holy war, 'far from being a pacifist or even a necessarily pacific religion, Christianity was from the beginning of the fifth century well prepared to act as the religion of societies that accepted war as a social duty'.[10] Familiarity with wars that had both secular and religious goals was widespread well before the eleventh century.

The early eighth-century account by the Venerable Bede of how King Oswald inspired his troops before they fought neighbouring pagan Britons is almost a preview of Urban II's famous call to take up the cross in 1095. As Bede put it, 'Oswald, moved by his devotion to the Faith, set up the standard of the holy cross before giving battle to his relentless enemies.'[11] Charlemagne's long and bloody campaigns against the Saxons around 780 had both a territorial and a religious goal: the expansion of the Carolingian realm and the introduction of the symbol of the cross to Saxony through the forced baptism of the Saxons.

In the last half of the ninth century, the effort to defend against the Scandinavian pagans who were ravaging Christian Europe's monasteries and churches was certainly motivated in part by religious consider-ations. Hence, when Alfred the Great made peace with the Danes at Chippenham in 878, he demanded that they convert to Christianity.

Similarly, after Charles the Simple and Rollo had made peace in Normandy around 911, Rollo and his Northmen converted to Christianity. A religious component is even more obvious in the defence against the attacks by Muslims and Magyars in the tenth and early eleventh centuries and in the Christian efforts to reconquer Spain.

Contrary to Hans Eberhard Mayer's emphasis on the role of that theological entity, 'the Church', in promoting the idea of the just war,[12] it is doubtful whether ecclesiastical ideas about just war had much influence on the reality of wars between 700 and 1050.[13] What is certain is that, as early as 700, warriors were becoming very familiar, not through preaching but from direct experience, with the fact that fighting could have religious as well as territorial goals. In fact, prior to the eleventh century there was little distinction between what we call Church and state, and consequently little distinction between secular and religious war, especially when it occurred at one of the physical frontiers of faith. But that changed dramatically with the changes in the distribution of power in Europe in the eleventh century.

By 1095, Western Europe was rapidly recovering, both economically and politically, from the damage caused by the Scandinavian, Muslim and Magyar invasions and was logistically able to go on the offensive. Moreover, and of central importance for our subject, dramatic religious developments during the eleventh century had brought a new and surprising degree of religious centralization to an otherwise very decentralized Europe. There was the spread of the first great centralized monastic order, the Cluniac Order,[14] and there was, above all, what is known as the Investiture Contest, the incomplete but startling success of papal efforts, between 1049 and 1122, to assert the independence of the papacy and its superiority over all Christians in Europe including kings.[15]

Prior to 1050, kings had exercised authority over both temporal and religious matters and fought wars from mixed religious and temporal motives – wars over which ecclesiastics had little if any control. After 1050, popes from Leo IX to Calixtus II began to claim that they alone had final authority over the use of military force in Europe. Not only that, the popes also increased their own direct involvement in the use of force by making some European kings and leaders their vassals (a tie that William the Conqueror refused), by intervening in a civil war in Germany, by sanctification of armies in Italy that fought to protect lands directly under the temporal control of the papacy and, in 1063, by Pope Alexander II's offer of remission of penance and absolution of their sins to those fighting Muslims in Spain.[16]

The great central figure in the Investiture Contest, however, was Gregory VII, pope from 1073 to 1085, and his pontificate might be considered the watershed between the earlier period of violence with mixed secular and religious motives and the new period in which unambiguously religious violence first appeared. As H. E. J. Cowdrey has put it in his monumental new biography of Gregory, 'Amongst the most striking features of Gregory's exercise of the papal office is the frequency and forcefulness with which he sought to recruit the laity of western Christendom, from kings and princes to the broad knightly classes, for one form or another of military service by placing their arms at the disposal of the apostolic see.'[17] And Cowdrey judges that, 'before all else, his motives were religious'.[18]

The precondition for major religious violence was nearly satisfied: popes were mustering the physical force that would enable them to attack people whom they considered religious enemies. And that possibility was strikingly realized on 27 November 1095, at the Council of Clermont-Ferrand. Urban II's military summons, the shout, 'God wills it', with which those assembled responded, and the military pilgrimage that followed, with a bishop representing the pope at its head, is the first and most famous instance of massive, unambiguously religious violence by European Christians.

In fact, four different types of religious violence were practised during the First Crusade. In the first place, in 1096, as a by-product of Urban's summons to holy war, some Christians perpetrated the first major massacres of Jews in Europe, mostly in northern France and the Rhineland.[19] These persecutors took off before the date set by the pope, they had no papal support for their actions, and they acted in defiance of local episcopal and imperial authority, but their violence was, nonetheless, indisputably religious. Whatever interest they had in booty, their explicit justification was that it was not the Muslims but the Jews close at hand who were the worst enemies of Christ; and during the assaults, they usually did not kill Jews who accepted baptism.

In the second place, there also occurred in 1096 what may be the most unambiguous and most extraordinary case of purely religious violence in medieval history. In the Rhineland, when the Jews who were attacked could no longer defend themselves, many used their remaining force on themselves for a purely religious reason: they committed collective suicide to sanctify the holy name and prevent the forced conversion of their children.[20]

The third type of religious violence was a long-lasting by-product of the massacres. During the first and subsequent crusades and other attacks, the persecutors not only killed Jews, they also forcibly converted some of them. But if the choice of baptism or death forced some Jews to do something most of them despised, it also presented ecclesiastical authorities with a problem. The doctrine prevailing since the Roman period prohibited the forcible conversion of Jews, and that doctrine would be reinforced around 1120 when, in reaction to the massacres of 1096, Pope Calixtus II issued the papal bull of protection for Jews that later popes would reissue at the time of crusades. The bull guaranteed Jews the right to live in peace and practise their religion in Europe, and it declared specifically 'that no Christian shall use violence to force them to be baptized as long as they are unwilling and refuse'.[21]

Moreover, when Jews were baptized by force, it was obvious that they were reluctant and that baptism had not suddenly changed them into believers in Christianity. Nonetheless, because of their theology, Catholic ecclesiastics had an issue to face. Were Jews who had accepted baptism because the only alternative was death really Christians, or were they still Jews and should therefore be allowed to practise Judaism? In fact, after the forced conversions of 1096, the German emperor and England's less than deeply devout king, William Rufus, recognized that the Jews were still Jews and wanted to remain so, and they allowed them to return to Judaism. But that apparently sensible and just resolution did not prevail.

The authority of the Catholic church rested on its control of the sacraments, as the Investiture Contest had just made very explicit. And although baptism could be administered by a lay person, it was defined by the Catholic church as a sacrament, the sacrament that freed people from original sin and made them members of the Catholic church and subject to its authority. Moreover, in contrast to the Mass, it was a sacrament whose effect was believed to depend neither on the priestly status of the person administering it, who could be a woman or a pagan, nor on the attitude of the recipient – who was typically an infant in the Middle Ages, and doubtless frequently a protesting infant at that. Its efficacy was believed to depend solely on the power of God. To declare the forced baptisms invalid would therefore have contradicted prevailing beliefs about the effect of the sacrament and denied God's power. Consequently, although Urban II apparently took no stand, subsequent popes declared forced baptisms binding, even though they had to engage in unbelievable quibbling to do so.

A letter by Innocent III in 1201, which became part of canon law, defined the issue in language that is worth quotation as an example of the ethical convolutions to which approval of religious violence could lead.

Truly, it is contrary to the Christian religion that someone who is always unwilling and wholly opposed to it should be compelled to adopt and observe Christianity. For this reason some distinguish, not absurdly, between unwilling and unwilling and between compelled and compelled so that he who is violently drawn into Christianity through fear and torture, and receives the sacrament of baptism in order to avoid loss, such a one (like one who comes to baptism in dissimulation) receives the impress of Christianity, and he, as conditionally willing though absolutely speaking unwilling, may be compelled to observe the Christian faith.[22]

A well-known and well-documented case of 1320 illustrates what that strange reasoning could mean in practice. In that year, what is known as the Shepherds' Crusade attacked Jews in southern France.[23] Some were killed and some converted out of fear. At Toulouse, Baruc, a learned German Jew, who saw other Jews being killed around him, accepted baptism under threat of death.[24] Thereafter, he returned to Pamiers, where he lived as a Jew in the Jewish community until he was denounced to the Inquisition as an apostate and brought before Jacques Fournier, the bishop of Pamiers, who later became Pope Benedict XII.[25]

Baruc claimed that he had never become a Christian. He described all the violence that had surrounded his baptism and said that, once safe, he had inquired of several priests whether a forced baptism was valid; and since they said that forced baptism was illegal, he had therefore continued in his Judaism. Fournier, however, did not accept Baruc's excuse and threatened him with death as an apostate unless he accepted to be a Christian. Baruc, apparently a learned man, replied that he would if he could be persuaded of the truth of Christianity. And here Fournier displayed the patience in bringing people back from what he considered error that he had demonstrated on several other occasions. The theological discussion stretched over eight weeks! Finally, Baruc, always under the threat of fiery death, gave in, received a new name after his abjuration, and became Jean Baruc.

Many other instances of forced conversion were the result, not of rioting mobs, but of official violence. The papal condemnation of the Talmud in 1242 falls somewhat outside my definition of religious violence because, although it involved the forceful seizure and burning of books, it did not call for physical attacks on Jews. The same might be said about the choice between baptism or expulsion in England in 1290 and in France in 1306, 1315 and 1380 which depended on the threat of physical enforcement. But many of the conversions in Italy in 1290,[26] in Spain between 1391 and 1492,[27] and in Portugal in 1497 were forced conversions resulting from mob violence.[28]

Religious violence against a Jew of an unusual kind could also occur because of the opposite situation. If Jews were killed because they had been baptized but had not really converted, conversely, a Christian who voluntarily converted to Judaism, a rare event, would be condemned to death, as an event at Oxford illustrates. It has been described with typical brilliance in an article by perhaps the greatest of English medievalists, Frederic William Maitland, whose attitude towards Jews was exemplary for an historian around 1900.[29]

In 1222, a deacon fell in love with a Jewish woman, accepted circumcision and voluntarily became a Jew. He was denounced to the authorities and condemned by a provincial council held at Oxford by Archbishop Langton. The day he was condemned was the second Sunday after Easter, which was, as Maitland noted, the day on which one sings in the introit 'The earth is full of the mercy of the Lord', and he was then burnt by royal officials. As Maitland put it with the sharp irony that pervades his article, 'There was no statute, there may perhaps have been no precedent to the point; such a case is not foreseen in advance; but that a deacon who turns Jew for the love of a Jewess shall be burnt, needed no proof whatever.'[30]

What all these instances demonstrate is that ecclesiastical authorities used religious violence by mobs and lay authorities to protect and enforce Catholic doctrine at the expense of Jews. They forced Jews to comply with judgements that flew in the face of their conscience and the empirical evidence, thereby tacitly approving *post factum* the prior illegal violence used against Jews by lay people.[31] The clerics most active in inciting this religious violence, particularly in Mediterranean Europe, were the Dominican and Franciscan friars,[32] and their persecutions had famous results. They produced the *conversos* or Marranos in Spain, the mass of genuine, half-hearted and unconvinced conversions produced by the religious violence of 1391 and by the order of expulsion from Spain of 1492.[33] The most famous result of their efforts, of course, was the Spanish Inquisition, which used torture and fire on allegedly hypocritical converts.[34]

During the First Crusade, there was also, of course, the fourth type of religious violence, the violence normal in holy wartime. The official crusaders, who started off later, did not massacre Jews in Europe, but from 1096 to 1099 they killed many Muslims as they fought their way towards Jerusalem. Not all those killings, however, can be categorized as religious violence by my restrictive definition, because many crusaders had very mixed motives. In addition to their religious goals, some of the leaders and their followers had rather naked political and territorial ambitions and were even sometimes reluctant to continue on to Jerusalem once they had acquired new possessions. But, especially below the knightly level, there were also those who were still fiercely motivated by their religious beliefs to capture Jerusalem and whose fighting, which involved highly gruesome brutality at times,[35] can certainly be categorized as religious violence.

The high point of this military violence came in 1099, at a highly emotional moment suffused with religious beliefs. After a religious procession around the walls of Jerusalem, and in the heat of battle after fighting their way into city, the crusaders waded knee-deep in blood, as the chronicler puts it, killing many of their Muslim religious enemies, doubtless including many who had given up all resistance, and also many Jews. The Jews had been defending their section of the city wall, and once the crusaders had got into the city, they also killed many Jews, including those killed when the crusaders burned down the synagogues or places of prayer to which Jews had fled. Many others were captured and sold as slaves. A similar massacre occurred later when Haifa was captured in 1100.[36] And within a few decades, the crusading commitment to religious violence would be institutionalized within the Catholic church itself in the military semi-monastic orders of the Knights Templar and the Hospitallers.

The physical frontiers of Christian Europe were now safe from invasion, but the triumph of the faith outside of Europe was short-lived. The Christians were soon in retreat in the Holy Land. In 1144, they lost Edessa; by 1187 they had lost Jerusalem; the following crusades were failures; and by 1297 all of the Holy Land was gone. Although the reconquest of Spain continued gradually and was nearly complete by the end of the thirteenth century, its territorial and religious motives were too mixed for it to count as religious violence in the restricted sense.[37] Similarly, the advance of Christianity in eastern Europe was accomplished by what Alfred Haverkamp has neatly termed 'a complex, inseparable symbiosis' of mission and conquest,[38] and it too cannot count as simply religious violence, although the ecclesiastical institutionalization of religious violence was represented by the activity of two military orders, the Knights of the Sword and the Teutonic Order.

In general, purely religious violence gradually declined outside of Europe after the eleventh century,[39] but inside Europe, a new and major chapter in the history of religious violence was just beginning. Because of educational recovery and the emergence and spread of independent religious thinking, new frontiers of faith were appearing within Europe itself. The great development of theological thinking which began in the eleventh century led some clerics and some independent-minded lay people to express doubts about some of the central beliefs of the Catholic church and to be condemned as heretics by the ecclesiastical authorities.[40]

In the eleventh and twelfth centuries, Church authorities condemned specific individual Christians as heretics, but the problem of deviant thought only increased, and the anger evoked by these deviations led to more institutionalization of religious violence within Europe. By the thirteenth century, popes were summoning crusades within Europe against collectivities of heretics or alleged heretics, the Cathars or Albigensians in southern France and the Stedingers in Germany; and a new papal institution, the Inquisition, was created which used the physical force of lay Catholics to extirpate deviant religious thought.[41]

The idea that Christians could, with ecclesiastical approval, use religious violence within Europe to control thought was now well established and heavily practised. But the most flagrant rejection of Catholic beliefs within Europe came not from those deemed heretics but from the open disbelief of Judaism and Jews, and the new atmosphere of enforced orthodoxy threatened them too. Thanks to the Christian doctrine established in the Roman period that justified toleration of Jews, and because of various demographic and political developments thereafter, by the late eleventh century the most advanced Judaic thought, such as that of Rashi, was to be found in various centres in Christian Europe. Here in the heart of Christianity, where the impressive explosion of Christian ecclesiastical organization, architecture and thought known as the Twelfth-Century Renaissance was occurring, Jews nonetheless remained confirmed – and open – disbelievers. They repeated, 'Hear Oh Israel, the Lord our God, the Lord is One', and thought of Jesus, not as god but as the hanged one, a putrid corpse, a dead body.[42]

To make the presence of Jewish disbelief more disturbing, by the late eleventh century, Jesus's death was acquiring a new and much more emotional meaning for an increasing number of Christians. Instead of thinking of events in the Gospels to a large extent as symbolic actions in the somewhat distant cosmic drama that was repeated liturgically in church every year, Christians were developing a new interest in the literal, historical meaning of the Gospels, an interest which the homecoming crusaders' stories about the Holy Land can only have intensified. There was a new recognition of the humanity of Jesus, a new emphasis on the details of his life and of his suffering and death[43] – and a new and more intense emotional reaction against the people Christians accused of being responsible for his death. In this new religious atmosphere, it would have been a miracle if many Christians had not been angered by the presence of Jews in their midst, and if they had not attacked Jews on various occasions.

Fortunately for Jews, they had gained some new protection. On the occasion of subsequent crusades, popes reissued the bull of protection that Calixtus II had inaugurated around 1120, and kings prohibited attacks on the Jews whom they profitably exploited. During the Second Crusade, Bernard of Clairvaux, whose letters in support of the crusade itself were a rhetorical peak in ecclesiastical appeals for religious violence, nonetheless used his immense prestige to stop a massacre in Germany.[44] The only major attack in England occurred in 1190 against the will of royal officials when Richard the Lionheart had just set off from England on the Third Crusade.[45]

The list of crusade attacks could go on and on, but it is uncertain what proportion of their motivation after the middle of the twelfth century was purely religious. Jews were excluded from most forms of commerce by then; they had become increasingly involved in, and had prospered from, the unpopular occupation of open moneylending; and many crusaders had to sacrifice and borrow to get the money they needed for their journey. Yet even though economic motives now played an important role in inciting attacks at the time of crusades, nonetheless, thanks to the increasing strength of royal and papal government in much of Europe, the number of victims in each of the subsequent crusade massacres was far less than in 1096.

But if unambiguously religious violence against Jews was declining, a new, deadly, and amazingly durable kind of accusation against Jews started around 1150. Although the new accusations did not cause significant violence before the middle of the thirteenth century, they proved deadly thereafter. This violence cannot, however, be categorized as religious violence against Jews in the normal sense because, although the accusations used religious language, they were not motivated by established Catholic beliefs about Jews and did not even refer to real Jewish conduct.

The hatred that had fuelled the massacre of 1096 was explicitly justified by the long-established Christian conviction that Jews did not believe in Christ and had killed him; a conviction that had a hard core of truth. In fact, not only had some Jews been involved in the death of Jesus, but since then, the Jews as a group, including those killed in

1096, denied that he was divine, believed that he had been properly condemned to death, and said derogatory things about him. The new type of violence differed radically. It was not motivated by long-established Christian convictions about Jews; it was impelled by totally false fantasies that had recently originated in the psyche of troubled believers.

The new accusations apparently arose because the unease and anger that the presence of the continuing disbelief of Jews in their midst inspired in some Christians was so out of proportion to anything the Jews were really doing that these disturbed Christians imagined that Jews were doing other much more horrible things in secret. They accused all Jews of engaging in a conspiracy to commit peculiarly horrible forms of physical violence against contemporary Christians, conduct that no one had in fact ever observed them doing. And they persuaded others who were susceptible to paranoia to believe their accusations and used these fantasies to incite physical attacks on Jews.

The first fantastic accusation was created in Norwich. In 1140, a married parish priest accused the Jews of killing William, his young nephew. There was no supporting evidence, but the cathedral chapter buried the alleged victim of Jews in the monks' cemetery; and when the monk, Thomas of Monmouth, joined the chapter around 1150, he created the fantasy that the Jews had crucified young William. According to Thomas, the Jews in Europe conspired every year to choose a country and a city by lot, and the Jews of that city would then kidnap and crucify a child as a ritual to free Jews from their exile. And in 1140, as all the Jews in England and Europe had known, the lot had fallen on the Jews of Norwich. Fortunately, many clerics in Norwich did not believe the fantasy, and neither did the sheriff, so no Jews were killed.[46] But the rumour spread, aided by the growing belief that Jews were murdering Christians in secret.

As a result, Jews in several cities in England and northern France were accused of crucifying Christians after a body was found whose killers were unknown. Yet despite the interest in these new saintly martyrs and the money their shrines brought in, and although the fantasy made all the Jews in Europe accomplices, Jews were only accused of the crime in northern Europe and only in a relatively few cities. Moreover, for a century, no Jews were killed because of that accusation.[47] Around the middle of the thirteenth century, however, the fantasy became lethal.

Jews were officially executed for alleged crucifixions in 1255 by Henry III for the death of young Hugh of Lincoln, in 1279 by Edward I for a body at Northampton, and in 1288 in France (to the annoyance of

Philip III) by local royal officials in cooperation with ecclesiastical authorities. In fact, relatively few Jews were killed in England and France because of the crucifixion accusation. It was in Germany that the great increase in victims took place. And it was in Germany that a new and different fantasy that was also not justified by Christian convictions appeared in 1235.

The Jews of the town of the great monastery of Fulda were accused of killing four boys in order to get the blood necessary for a Jewish ritual, and 24 Jews were slaughtered.[48] The fantasy got immediate and Europe-wide publicity because someone had the bodies transported across many kilometres to be presented to the emperor, Frederick II, with a demand for justice. Frederick did not believe the story and convoked converted Jews from across Europe. They declared the accusation ridiculous, and Fredrick then promulgated an imperial bull that declared the Jews innocent, denounced the falsity of the accusation, and prohibited any more such accusations. Nonetheless, there were more attacks, and 11 years later, Pope Innocent IV also condemned the ritual cannibalism accusation, although neither he nor other popes ever condemned the ritual crucifixion accusations.[49]

But much worse was about to happen. Another new and extremely deadly fantasy was born in France in 1290, but only led to widespread violence when it reached Germany. According to Catholic dogma, when priests consecrated the wafers used in the Mass, the wafers were transubstantiated into the real body of Christ. Although it had been taken for granted for centuries that Jews did not share that belief, nonetheless, the new fantasy accused Jews of stealing consecrated hosts in order to torture Christ. The fantasy accused the Jews of mistreating the host until it bled and cried out. The accusation was not justified by prevailing Christian convictions about Jews or by the reality of contemporary Jews but by the projection of Christian beliefs onto Jews, by the totally false assumption that the Jews themselves believed that the host had become Christ's real body, and that they could therefore torture it anew.

It was a peculiarly lethal accusation. Ritual murder accusations had remained limited in number and location because they required a dead body, but wafers could be discovered everywhere. The new fantasy enabled people to believe that Jews were torturing Jesus in many places at the same time, and they were encouraged so to believe by the friars who spread stories about the cries of the tortured hosts to prove that Christ was really present in the host.[50] In 1298, when Jews in Röttingen were accused of torturing the host, the accusation sparked widespread massacres in Germany in which at least 3400 and possibly between

4000 and 5000 Jews were killed. The same accusation in 1336 led to the Armleder movement, which may have slaughtered 6000 German Jews.[51] Many more Jews were killed by these massacres than during the crusades.[52]

A very different charge swept Europe between 1348 and 1350. The Jews were accused of poisoning the wells and causing the Black Death, and thousands of Jews were killed. These massacres too were not instances of unambiguously religious violence, for they were not justified by Christian doctrine about Jews; indeed the pope declared the accusation false. Rather, they were an obvious case of scapegoating an unpopular group for a major disaster. Nor were the expulsions of Jews from almost all of western Europe by the end of the fifteenth century justified by Catholic doctrine. If anything, they ran against the spirit of the papal bull of protection for Jews.

The unreality of the fantastic accusations is strikingly reflected in the ambivalent way that popes handled them. On the whole, the papacy was surprisingly silent about these fantasies. It did condemn the ritual cannibalism accusation, which had no Christian religious significance, other than its projection on Jews of the Christian belief in the necessity of blood for salvation, and it condemned the Black Death fantasy which also had no significance for Christian belief. But it did not condemn the ritual murder by crucifixion fantasy, for it reinforced the belief that Jews were crucifying murderers. And it did not condemn the fantasy about torturing the consecrated host of the Mass, for it reinforced the dogma that Christ was really present in the wafer. Rather, by accepting the ongoing establishment of local shrines to honour the alleged victims of the Jews, whether tortured hosts or humans, the papacy tacitly supported the accusations. Indeed, in the famous case of Simon of Trent, it managed to avoid supporting the fantasy itself while nonetheless approving the piety of those who believed it.[53] Apparently, the papacy itself did not consider that the content of these accusations was the product of religious belief.

What is most obvious from this brief survey is that the violence inflicted on Jews changed radically in the course of the Middle Ages. From 500 to 1096, there was little violence. Then came the great outburst of violence in 1096. Although the violence that accompanied the First Crusade and, to a lesser extent, the subsequent crusades, may have been contrary to the rather abstract and highly ambiguous Catholic doctrine that Jews should be tolerated, it was indisputably religious violence. It was an aggressive overreaction to an ecclesiastical summons to holy war against the enemies of the faith, and it was clearly motivated by specific beliefs about Jews that had been firmly established in

Catholic doctrine for a very long time. What is interesting to note is that widespread violence against Jews that was indisputably religious in character did not last. It only began in 1096 and was already declining in the twelfth and thirteenth centuries.

Offsetting that decline, however, was the appearance of a very different kind of violence. Starting around 1250, Jews became the targets of pyschopathological violence, as they would continue to be down to the twentieth century. By psychopathological violence, I mean violence that was motivated and explicitly justified by the irrational fantasies of paranoid people whose internal frontiers of faith were threatened by doubts they did not admit. The violence incited by accusations of ritual murder, torture of the host and well-poisoning might seem to be religious violence against Jews because the accusations used religious language and talked about Jews, but it was not. The violence was a psychological defence against an imaginary enemy.

The pathological violence of these Christians sprang from paranoia and doubt, not from belief. Although the paranoiacs and those who believed and followed them accused and killed the people they called Jews, they were not thinking of real Jews. They were projecting imaginary evil characteristics on Jews in an irrational effort to bolster their confidence in the value of their own identity as Christians. By their accusations and violence, they expressed their need to believe that Jews were inhuman murderers both of God and of contemporary Christians, whereas they, by contrast, were righteous and would be saved by Christ, whose support was comfortingly close in the consecrated host of the Mass.

Jews were still, of course, the object also of anti-Judaic hostility and the usual hostility against moneylenders, so that the mixture of motives involved in the late medieval attacks is almost impossible to disentangle. By the end of the Middle Ages, Jews had become, especially in northern Europe, an all-purpose bogyman for those Europeans who needed an outlet for their paranoia.[54] And, partly in reaction to those fears, Christians constructed ghettos and expelled Jews from most of western Europe, thereby closing off or eliminating the physical frontiers where their faith might have to confront the reality of Jews and Judaism. But they could not eliminate awareness of the mental frontiers because the millennial challenge of Jewish disbelief in Christ was enshrined in the New Testament and deeply embedded not only in their religion but also in their art and literature. Consequently, consciousness of that disturbing frontier of faith remained even when or where Jews were physically absent.

154   *Gavin I. Langmuir*

Notes

1. For a rich discussion of the violence connected with monotheism in the Bible, primarily in the Hebrew Scriptures, see R. M. Schwarz, *The Curse of Cain: The Violent Legacy of Monotheism* (Chicago, IL,1997).
2. See P. Badham (ed.), *A John Hick Reader* (Basingstoke, 1990).
3. The paper in which I first argued this position, 'Intolerance and Tolerance: Only One "One and Only" God or More', was delivered in September 1996 at the conference in honour of the centenary of the birth of James Parkes, but the proceedings of the conference have not yet been published.
4. For a discussion in depth of the relation between religion viewed empirically and hostility to Jews, see my *History, Religion, and Antisemitism* (Berkeley and Los Angeles, CA, 1990), and 'The Faith of Christians and Hostility to Jews', in D. Wood (ed.), *Christianity and Judaism*, Studies in Church History, 29 (Oxford, 1992) pp. 77–92.
5. For the difference between Islamic and Christian attitudes towards Jews, see Mark R. Cohen, *Under Crescent and Cross: The Jews in the Middle Ages* (Princeton, NJ, 1994).
6. See Langmuir, *History, Religion, and Antisemitism*, pp. 248–51, 259–61.
7. The classic study of the earlier period is B. Blumenkranz, *Juifs et Chrétiens dans le monde occidental, 430–1096* (Paris, 1960).
8. See Langmuir, 'The Faith of Christians and Hostility to Jews', above n. 4.
9. The policy of forced conversion and the violence of the Catholic Visigothic kings in Spain does not seem a case of religious violence because there was such a mixture of political and religious motives involved.
10. P. Partner, *God of Battles: Holy Wars of Christianity and Islam* (Princeton, NJ, 1998) p. 29.
11. Bede, *A History of the English Church and People*, trans. L. Sherley-Price (Harmondsworth, 1956), Bk 3, ch. 2, p. 140.
12. H. H. Mayer, *The Crusades*, 2nd edn (Oxford, 1988) pp. 14–15. Mayer does think, however, 'that too much attention has been paid to the eleventh-century developments in the Church's concept of a holy war' (ibid., p. 20).
13. Guibert of Nogent wrote in his *Gesta* about 1110 that 'wars traditionally have been fought absolutely legitimately only for the protection of Holy Church. But because nobody has had this right intention and the lust for possessions has pervaded the hearts of all, God has instituted in our time holy wars' (quoted in J. Riley-Smith, *The First Crusade and the Idea of Crusading* (London, 1986) p. 144).
14. For a brief description, see C. H. Lawrence, *Medieval Monasticism*, 2nd edn (London, 1989) pp. 86–102.
15. For an excellent collection of documents in translation relevant to that much-discussed conflict, together with an excellent commentary, see B. Tierney, *The Crisis of Church and State* (Englewood Cliffs, NJ, 1964) and for a recent short description of it, see U.-R. Blumenthal, *The Investiture Controversy: Church and Monarchy from the Ninth to the Twelfth Century* (Philadelphia, PA, 1988).
16. See Riley-Smith, *The First Crusade and the Idea of Crusading*, p. 5.
17. H. E. J. Cowdrey, *Pope Gregory VII, 1073–1085* (Oxford, 1998) p. 650.
18. Ibid., p. 695.

19. See R. Chazan, *European Jewry and the First Crusade* (Berkeley and Los Angeles, CA, 1987).
20. Ibid., pp. 99–136; F. Lotter, '"Tod oder Taufe"; das Problem der Zwangtaufen während des Ersten Kreuzzugs', in A. Haverkamp (ed.), *Juden und Christen zur zeit der Kreuzzuge*, Vorträge und Forschungen, 47 (Thorbecke, 1999) pp. 107–52.
21. See S. Grayzel's excellent collection of translated documents, *The Church and the Jews in the XIIIth Century*, vol. 1, rev. edn (New York, 1966) pp. 76–83 and 92–5.
22. Ibid., pp. 100–3; this is the version issued by Innocent III in 1199.
23. See D. Nirenberg, *Communities of Violence: Persecution of Minorities in the Middle Ages* (Princeton, NJ, 1996) pp. 43–6.
24. See A. Pales-Gobillard, 'L'Inquisition et les Juifs: Le cas de Jacques Fournier', in M. H.Vicaire (ed.), *Juifs et judaïsme de Languedoc*, Cahiers de Fanjeaux, 12 (Toulouse, 1977) pp. 97–114.
25. On how Fournier and the Inquisition in Languedoc operated, see J. B. Given, *Inquisition and Medieval Society: Power, Discipline and Resistance in Languedoc* (Ithaca, NY, 1997).
26. See A. Milano, *Storia degli ebrei in Italia* (Turin, 1963) pp. 101–4.
27. See A. Foa, *Ebrei in Europa dalla pest nera all'emancipazione* (Rome, 1992) p. 111, where she states that the motivation of the violence in 1391 was primarily religious, not social or economic. See also H. Kamen, *The Spanish Inquisition: A Historical Revision* (New Haven, CT, 1998) pp. 10, 12–16.
28. For an overview of the major instances of forced conversion with a vast bibliography about them, see S. T. Katz, *The Holocaust in Historical Context*, vol. 1: *The Holocaust and Mass Death before the Modern Age* (Oxford, 1994) pp. 362–75.
29. See G. Langmuir, *Toward a Definition of Antisemitism* (Berkeley and Los Angeles, CA, 1990) pp. 32–5.
30. 'The Deacon and the Jewess: or Apostasy at Common Law', in H. A. L. Fisher (ed.), *The Collected Papers of Frederic William Maitland* (Cambridge, 1911) pp. 153–65. For another instance of the execution of ex-Christians, see Kamen, *The Spanish Inquisition*, p. 19.
31. And those attitudes endured. They could still be found among Catholic ecclesiastics after the Holocaust, as the Anneke Beekman affair in the Netherlands demonstrated; see J. S. Fishman, 'The Anneke Beekman Affair and the Dutch News Media', *Jewish Social Studies*, **40** (1978) 3–24.
32. For a valuable, if somewhat exaggerated, analysis of their role, see J. Cohen, *The Friars and the Jews: The Evolution of Medieval Anti-Judaism* (Ithaca, NY, 1982).
33. See Y. Baer, *A History of the Jews in Christian Spain*, vol. 2 (Philadelphia, PA, 1966) pp. 95–134.
34. See Kamen, *The Spanish Inquisition*, pp. 17–65.
35. See R. C. Finucane, *Soldiers of Faith: Crusaders and Moslems at War* (New York, 1983) pp. 100–3.
36. See J. Prawer, *The History of the Jews in the Latin Kingdom of Jerusalem* (Oxford, 1988) pp. 19–40.
37. See Kamen, *The Spanish Inquisition*, pp. 1–2.
38. A. Haverkamp, *Medieval Germany, 1056–1273*, trans. H. Braun and R. Mortimer (Oxford, 1988) p. 17.

39. Even though awareness of how extensive the challenge of Islam was had increased (see Langmuir, *Toward a Definition of Antisemitism*, ch. 8).

40. See R. I. Moore, *The Origins of European Dissent* (New York, 1977) and M. Lambert, *Medieval Heresy*, 2nd edn (Oxford, 1992).

41. See, amongst many other possibilities, R. Kieckhefer, *Repression of Heresy in Medieval Germany* (Philadelphia, PA, 1979).

42. J. Katz, *Exclusiveness and Tolerance: Jewish–Gentile Relations in Medieval and Modern Times* ([London], 1961) pp. 22–3; Chazan, *European Jewry and the First Crusade* pp. 67, 108, 129.

43. See Langmuir, *Toward a Definition of Antisemitism*, pp. 76–8, 115–16.

44. See Katz, *The Holocaust and Mass Death*, pp. 331–2.

45. See R. B. Dobson, *The Jews of York and the Massacre of March 1190*, Borthwick Papers, no. 45 (York, 1974).

46. Langmuir; 'Thomas of Monmouth: Detector of Ritual Murder', in *Toward a Definition of Antisemitism*, ch. 9.

47. There were accusations in about 25 towns in northern Europe by 1350: Langmuir, 'L'absence d'accusation de meurtre rituel à l'ouest du Rhône', in M. H. Vicaire and B. Blumenkranz (eds), *Juifs et judaïsme de Languedoc xiie siècle–début xive siècle*, Cahiers de Fanjeaux, 12 (Toulouse, 1977) pp. 240–2.

48. Langmuir; 'Ritual Cannibalism', *Toward a Definition of Antisemitism*, ch. 11.

49. Grayzel, *The Church and the Jews*, pp. 268–71.

50. Langmuir, 'The Tortures of the body of Christ', in S. L. Waugh and P. D. Diehl (eds), *Christendom and its Discontents* (Cambridge, 1996) pp. 287–309; and for a rich collection of accusations of this kind, see now M. Rubin, *Gentile Tales* (New Haven, CT, 1999).

51. See F. Lotter, 'Die Judenverfolgung des "König Rintfleisch in Franken um 1298"', *Zeitschrift für historische Forschung*, **15** (1988), 385–422; and idem, 'Hostienfrevelvorwurf und Blutwunderfalschung bei den Judenverfolgung von 1298 ("Rintfleisch") und 1336–1338 ("Armleder")', in *Fälschungen im Mittelalter*, MGH, Schriften, vol. 33, part 5 (Hanover, 1988) pp. 533–83.

52. See Katz, *The Holocaust and Mass Death*, pp. 334–5.

53. For a full description of the events see R. Po-Chia, *Trent 1475* (New Haven, CT, 1992); for a neat summary of the papal position, see Foa, *Ebrei in Europa*, p. 50.

54. For an outstanding short survey of antisemitism through the ages, see M. Shain, *Antisemitism* (London, 1998).

8
Christians and Jews and the 'Dialogue of Violence' in Late Imperial Russia*

John D. Klier

When the first sounds of breaking glass signalled the onset of an anti-Jewish riot, a pogrom, in Odessa, Christians hastened to place crosses and icons in the windows of their shops and homes, confident that they would provide protection from attack. Christian servants in Jewish homes which were ransacked were able to save themselves from a beating by making the sign of the cross. (Jews accosted in the street by the mob were also sometimes forced to perform what was for them a humiliating gesture.) Russian authorities intent on maintaining order in the street, as well as potential Jewish victims of violence, linked the danger of pogroms to Christian religious feast-days, particularly the celebration of Russian Orthodox Easter. In the light of these well-confirmed facts, how can it be denied that pogroms in the Russian Empire grew out of religious prejudice and hate?

I have in fact argued elsewhere (most recently in *The Jewish Quarterly*[1]) against the utility of the concept of 'traditional Russian religious Anti-semitism' as an interpretative framework to understand the policies of the Russian government towards the Jews in the modern period. In this essay, I would like to extend this claim to question the primacy of religious prejudice as the prime ingredient in provoking the pogroms. While I do not deny the role of religious difference in creating the idea of the Jew as an alien and an 'other' in popular attitudes, I do make the case that other considerations were at play, overriding popular prejudice, such as the view of the Jews as 'the enemies of Christ'. In short, I contend that Christians and Jews engaged in a daily 'dialogue of violence' within the Russian Empire, which only took the form of pogroms on rare occasions, usually at moments of the collapse of central control and order. The actual characteristics of Russian pogroms, as they unfolded, cannot be attributed purely to religious factors.

This essay focuses on the pogroms of 1881–2 for a number of reasons. They were the first mass outbreak of anti-Jewish violence in the Russian Empire.[2] Their number and diversity, and the archival documentation now available on them, provides copious evidence that can assist their proper characterization. Over 250 events took place in the Russian Empire in 1881–2 which were classified as 'pogroms' and helped to popularize the name.[3] There was no common pattern. The pogroms of 1881–2 included the three-day Bacchanalia in Kiev, with several fatalities and massive property damage, as well as numerous incidents, where several peasants broke tavern windows. As one Russian governor correctly noted, the climate of fear and concern lent special significance to events which in normal times would have passed unremarked.

The pogroms of 1881–2 provided a precedent and a model that later pogroms would follow, at least in their early stages. The first pogrom wave provoked a good deal of attention from contemporaries, official and unofficial, who sought to explain the pogrom phenomenon. *De facto*, these pogroms made mass anti-Jewish violence an accepted part of the culture of late imperial Russia, prone to occur in times of crisis and the breakdown of state authority.

This latter statement is a starting point from which quickly to dispose of a number of myths which have grown up around the Russian pogroms. First, they were not the result of conscious governmental policy.[4] In 1881–4 (although less so in 1903–7), the Russian state did everything in its power to forestall, anticipate and prevent pogroms. It acted vigorously to suppress them when they broke out, and sought to investigate them thoroughly and punish the instigators and perpetrators, especially local officials who were thought to be remiss in their duties.[5] In 1903–7, pogroms took place when the state power was weak and under threat from revolution. They were part of a general breakdown of order, albeit with their own specificities. In 1919–21, in the midst of civil war, the imperial Russian state had ceased to exist at all. There is abundant evidence that pogroms were unplanned. They might best be seen as outbreaks of mass hysteria, spread as copycat violence. There was no guiding ideology which underlay the pogroms.

Russian pogroms should be set in the context of violence in Russia as a whole, for while pogrom-style violence was common, Jewish pogroms were not. In the second half of the nineteenth century alone, the Russian state had to contend with mass violence by peasants and workers, the so-called cholera riots (1831), the vodka riots (1861), urban disorders directed against the police (Rostov-na-Donu, 1879) and interethnic rioting, with religious overtones, in the Caucasus (Baku, 1881). To this list must be added the rising revolutionary movement and violence prompted by the struggle of national movements (Poland, Finland, Armenia et al.) against Tsarist rule. In short, violence in Russia was always threatening to break out, and many agencies of tsarist rule existed precisely to ensure that it did not. The significant questions which must be answered for 1881–2 are: why did violence occur, why was it directed against the Jews, and why did it take the forms that it did?

The pogroms occurred against the background of the uncertainty following the assassination of Emperor Alexander II on 1 March 1881.[6] I have challenged the claim that the Russian press conducted a sustained campaign against the Jews, implicating them in the assassination because of the involvement of Ges'ia Gel'fman. There is certainly no truth to the claim that the press instigated violence against the Jews. (A number of press organs did routinely print Judeophobe attacks on the Jews, and this gave rise to the legend of press instigation.) Some *pogrom-shchiki* were reliably reported to have shouted that they were avenging the Tsar's death. The claim that the Jews were regicides must have travelled as rumour, part of the fantastic assemblage of such stories that circulated throughout Russia after 1881. The Judeophobe claim that the Jews were targeted because they were guilty of 'exploiting the people' also lacks credibility, although it was widely believed by high officials.[7]

The general setting for the outbreak of the pogroms was the celebration of Easter; the immediate occasion was an episode of the everyday antagonisms which existed between Jews and their neighbours – first as members of different social classes and secondarily as a community of religious 'Others'. It is useful to recall that the Russian Empire was divided along ethnic, religious and class lines, and antagonisms could arise from either a single element of difference, or from a combination.

Another ingredient, which I shall term 'cultural', should also be noted. Those unfamiliar with the celebration of Russian Easter have often tended to see it as a time of heightened religious excitement. To some extent it was, but just as important were accompanying feelings of liberation and carnival. Easter concluded a lengthy period of denial – the Great Fast. It ushered in a period known as 'Bright Week', which was marked by boisterous celebration, grand meals and public entertainments. A central feature of the celebrations, frequently lamented by the devout, was wild intoxication and revelry. It was drunken aggression, as much if not more than religious devotion, which fuelled holiday celebrations. Russian officials recognized this danger by strengthening the police and assigning military patrols in areas where clashes between Christians and Jews might be a possibility.[8] Indeed, troops had just been stood down from their holiday patrols when the first pogrom broke out in the Ukrainian town of Elizavetgrad, a major trade and railroad centre, on 15 April 1881.

The locus of the outbreak was the tavern of the Jew Shulim Grichevskii. One account claimed that a local 'holy fool', Ivanushka, was manhandled for breaking a three-kopeck glass, another that the pretext was his noisy singing of the Easter anthem 'Khristos Voskres' ('Christ is Risen'). In any event, his cries attracted a large crowd. A policeman arrived, and urged the crowd to disperse. Instead, they began to shout 'The Yids are beating our people!' and to attack Jews. The mob proceeded to Bazaar Square, breaking shop windows and throwing goods into the street. Jewish shopkeepers defended themselves with crowbars and axes. The rioting moved on to nearby Jewish homes, and surrounding streets filled up with broken furniture. The contents of ripped featherbeds created the surreal illusion of a sudden snowfall. Troops were recalled to the city, but it was too late. The riot continued until the next day.[9]

Significantly, the next large incident broke out in a town far away from Elizavetgrad, but connected to it by a direct rail link. The copycat violence that characterized 1881 spread out along lines of communication such as railroads, canals and main highways.[10] Meanwhile, smaller incidents, chiefly attacks on isolated Jewish taverns, were carried out by peasants who were returning home from market towns where they had witnessed pogrom violence. Some peasants actually headed for towns in order to participate in looting. Official announcements condemning the pogroms, as well as newspaper accounts of events, also spread the pogrom model.

It is important to differentiate the various sorts of pogroms into at least two broad categories. There were the large-scale urban riots, which required the intervention of the army, caused millions of roubles worth of damage, and occasionally resulted in the deaths of both Jews and *pogromshchiki*, as the rioters were styled.[11] There continues to be scholarly disagreement over the identity of these urban rioters.[12] Rural 'pogroms', in contrast, were small-scale, hardly ever involved violence against persons (although isolated rural Jews were theoretically much more vulnerable), and were often over by the time the authorities arrived. The perpetrators were almost always peasants or persons who had ties to the countryside. I would emphasize the small-scale, copycat nature of these events, to suggest that the urban pogroms are of greater

interest for a proper understanding of the pogroms. Occasional reference to salient features of rural pogroms will nonetheless be made as
well.

Let us return to the immediate question of 'why the Jews?' In the
unsettled conditions of the summer of 1881, violence was close to the
surface, and any pretext might have triggered it off. It is noteworthy
that a pogrom erupted in Baku on exactly the same day as the Elisavetgrad disorders. These two pogroms were almost identical, except that
the Baku events involved a clash between Muslims and Christians.[13]
While taking preventive measures to anticipate anti-Jewish disorders,
the officials also expressed the fear that violence might break out
against the landowning nobility, whom rumour also blamed for killing
the Tsar, because he wished to give more land to the peasants.[14] In other
words, any outsider group – whether defined in social/class, religious or
ethnic terms – was vulnerable to attack. Thus, it would be a mistake to
view the attacks on Jews in isolation from the wider context of violence
within the Russian Empire, and still more misleading to attribute it to
religious hostilities alone.[15] Yet there were distinctive aspects to the
anti-Jewish pogroms, which cast light on the mutual relations of Jews
and non-Jews.

Historiography has been so inclined to view the Jews of the Russian
Empire as a persecuted religious minority that it has been blind to issues
of status and social hierarchy. These must be recognized, especially at a
time when they were overturned in a period of role-reversal or 'carnival', such as characterized the first pogroms. Russian Judeophobes, in
contrast, were fond of depicting the Jews of the Empire as a privileged
minority, who had access to special rights and prerogatives unavailable
to the masses.[16] What was the status of the Jews in the Russian Empire,
both *de jure* and *de facto*?

The Jews of the Russian Empire were members of a tolerated faith
community,[17] with extensive rights of internal administration for the
religious and social life of the community. While the Russian state left
internal life to the Jews themselves, their socio-economic life in the
wider community was subject to an enormous corpus of regulatory law.
Much of this law – although not all – circumscribed Jewish economic
pursuits. The most notorious body of law created the so-called 'Pale of
Settlement', residential restrictions on where in the Russian Empire
Jews might live. On a number of occasions, most notably 1804 and
1835, sweeping codes of law, based on general principles, had been
introduced for Jews. More commonly, law was introduced on an ad
hoc basis, responding to concerns of the moment. This produced

inconsistency, contradiction and ample scope for official caprice. Two examples may suffice.

In general, the Russian state was eager to move the Jews away from petty trade and into agriculture, and as early as 1804 it promoted agricultural colonization and rural settlement of Jews. It initially welcomed the investment of Jewish capital in land. Yet almost simultaneously, in response to a perceived threat of 'Jewish exploitation of the peasantry', and fear of Jewish links to Polish landowners in the strategic western borderlands, the state sought to restrict Jewish settlement in the midst of the peasantry, and to obstruct the lease or purchase of land by Jews.[18]

Another example is provided by the tavern trade. The regime fought a century-long battle against the concentration of Jews in the production and sale of spirits within the Pale of Settlement. Yet it could never bring itself to impose a complete ban, and even allowed Jewish distillers and brewers to enter the Russian interior. This indulgence was directly linked to the economic significance of the state vodka monopoly.[19]

The contradictions of official policy meant that circumvention of the law was a daily reality. Large sections of Kiev, located in the very heart of the Pale, were off-limits for Jewish settlement. Given the size of the surrounding Jewish population, the economic opportunities offered by Kiev, and the willingness of non-Jews to violate the law by renting to Jews, a large, illegal Jewish settlement was to be found in Kiev. At regular intervals the authorities would correct the situation by conducting police raids, a 'hunt' (*oblava*), which rousted out hundreds of illegal settlers. These miscreants would be marched out of the city under armed guard, sometimes in chains. The following day, the process of resettlement would begin until the next 'hunt'. When the regime decided to crack down on Jewish settlement outside the Pale, as in Moscow in 1891, literally thousands of 'illegal' families were found in residence, some of decades-long standing. It has been the 'hunts', pogroms and the Pale which have usually attracted the attention of historians of Russian Jewry. Yet an emphasis on violence and persecution obscures other, more positive, aspects of Jewish life in the Empire. This is especially so if they are placed in comparative perspective.

First and foremost, the Jews enjoyed personal freedom. This was in contrast to the majority of the Russian rural population who were serfs before 1861, and bound juridically to the community and their place of residence even after emancipation. Within the Pale, the Jews were linked to the land-holding nobility (the Polish *szlachta*), serving on their estates as managers, agents and leaseholders of the numerous economic prerogatives (monopolies on the use of woods, fishponds, mills) that accrued to the owner. The Jews served as the principal link between the countryside and the market, and the peasantry was almost completely dependent upon them in this regard. Reflecting their origins in Poland as an urban people, Jews were also highly concentrated in handicrafts. All surveys concluded that 'crafts and the marketplace are completely in their hands'. To this should be added the village tavern, the stereotypical 'Jewish space' in the countryside. Jews in the Pale were to be found almost equally divided into rural Jews (living in the countryside and in market towns, the famous *shtetlakh*) and urban Jews resident in the larger towns of the western provinces (some of which had grown up from being *shtetlakh*).[20] Jews were also an important component of cities, such as Odessa, Lodz and Warsaw, which underwent rapid growth into major commercial centres in the second half of the nineteenth century.[21]

The marketplace was the quintessential meeting place for Jews and non-Jews, and it was an environment where the Jews felt confident and at home. Like markets around the world, it was also a centre for disagreements, insults and fights. Jewish stall-holders felt no compunction against trading insults with Christian competitors, importuning potential buyers, manhandling troublesome customers or boxing the ears of the street urchins who filled the marketplace.[22] Tavern-keepers, whose livelihood depended on catering to human weakness, had even less respect for many of their customers, especially those who asked for credit or became drunk and disorderly. Such patrons were unceremoniously shown the door. In short, the meek and mild Jew, cringing before the Gentiles, is very much a fictional creation.

The confidence of Jews *vis-à-vis* their Christian neighbours was equally on display in *shtetlakh* and the quarters of towns where Jews constituted a large percentage of the population. Christians complained of the noise from Jewish synagogues, or the rowdy aspects of Jewish weddings, held in the open air. Jewish influence was most overtly demonstrated on the Sabbath, when the economic life of the villages came to a complete halt. This was in contrast to Sundays, when peasants emerging from church would find Jewish stalls and taverns open and ready to serve their customers. At least some were ready to take offence.[23] Another target for Christian resentment was the *eruv*, or ritual boundary, which surrounded many a *shtetl* and emphasized its Jewish nature.[24]

The proximity of Christian and Jew, and the public nature of at least some of their worship, generated curiosity, but did not promote understanding. Peasants in one small village, a newspaper article claimed, surrounded the Jewish synagogue on Yom Kippur, in order to test the local legend that the devil appeared in the synagogue to carry off the greatest sinner. The peasants carefully noted who emerged from the service. In Odessa, by contrast, urchins from the Jewish quarter gathered on Easter eve outside the nearby Greek church to watch – not always respectfully – the solemn procession around the church, the 'Khrestyi khod', which was a dramatic feature of the Pascal service. Fights and stone-throwing were a common feature of the season in Odessa, involving Jewish and Christian youths in what might best be described as 'turf wars'. It is easy to understand why the Odessa pogrom of 1859 was prompted by a rumour that Jews had desecrated a cross outside the Greek church. Again, at first glance this might appear a religious dispute, but it might better be seen as the contesting of space. It is interesting that the first official reports on the pogroms of 1881 often characterized them, not as 'attacks upon the Jews', but as 'clashes between Jews and Christians'. Indeed, one of the most common recommendations for the prevention of pogroms, made by officialdom in 1881, was that 'the Jews should conduct themselves more modestly'.[25]

Let us return to Bright Week in 1881. It should have been the culmination of the liturgical year for Orthodox Christians with feasting, drinking and public celebrations – in short, a form of carnival. Instead, the traditional street fairs, whose rides attracted a large public, were muted by the authorities because of the mourning period for the assassinated emperor. Jews in some areas allegedly mocked Christians with the claim that 'we have bought your holiday!' When troops arrived at the start of Bright Week as a holiday precaution, it was claimed that Jews taunted their neighbours that they had best be on their good behaviour, because the authorities were protecting them. When *pogromshchiki* were arrested in the first hours of the Elisavetgrad pogrom, Jews were alleged to have mocked a column of prisoners with shouts of 'they'll string you up like dogs!'[26] *Pogromshchiki* often defended themselves with the claim that Jewish provocation triggered events. (Incidentally, it does not matter whether such provocation actually took place – what is important is that they unquestionably constituted a large body of rumour which had the effect of inflaming the population.)

The cry of *'bei zhidov!'* ('beat the Yids') was to become the archetypal slogan of organized and semi-organized violence against Jews at the start of the twentieth century, and the symbol of the pogrom-mongering 'Black Hundreds'. The shout was occasionally heard in 1881. But just as common was the cry, heard outside Grichevskii's tavern, that triggered the first pogrom in Elisavetgrad: *'zhidy b'iut nashikh!'* ('the Yids are beating our people!') From what has been said above, it was not an uncommon event for Christians and Jews to engage in fisticuffs. Indeed, the Jews were not the passive victims of pogrom mythology. Far from running away, the Jews of Elisavetgrad initially defended their shops with crowbars and axes until overwhelmed by superior numbers. Throughout 1881, Jewish communities organized self-defence units, sometimes with the approval of the authorities. At the onset of the notorious Balta pogrom during the Pascal season of 1882, the rioters gained the upper hand only after the authorities had disarmed and dispersed a large crowd of Jews on the central square, who had gathered to defend their property.[27] In other words, Jewish resistance was a familiar part of pogroms.

The pogroms themselves seem to have largely followed a set of unwritten rules. They were directed against Jewish property only. The initial objective of the pogrom was the physical destruction of Jewish property, not looting. (This was carried out by the pogrom camp-followers, often peasants and women.) The most important exception was the looting of taverns, where the stock was invariably consumed, often with fatal consequences. (A number of pogrom deaths were caused by alcohol poisoning.) Besides the obvious fact that alcohol was the most easily looted commodity, it also allowed the crowd to vent its frustration on an institution, the state-licensed tavern, which was the target of popular antipathy.[28]

Most contemporaries claimed that the pogroms were directed against property and not against persons. This claim is hard to measure, although given the scale of the pogroms there were remarkably few deaths. Likewise, official attempts to confirm foreign atrocity reports suggested that they were exaggerated. A more ticklish question is that of rape, a notoriously under-reported crime. The foreign press was filled with gruesome reports of rape on a massive scale, at least one case of which, in Balta, was investigated by the authorities.[29] On the whole, there is little evidence in the archives of rape as a common occurrence. Ironically, a higher incidence of rape might buttress the hypothesis that religion was not the chief motivation of the pogroms. Modern scholarship suggests that rape in situations like the pogroms can be viewed as an assertion of power.[30] In this case, it would have conformed to the idea of reversal of status, of 'showing the Jews who is boss'. This is clearly an area which requires more investigation.

If Jewish resistance was an accepted part of the 'rules' of a pogrom, this would explain the absence of fatalities or serious injuries: pogrom events were part of the everyday violence of Russian life. There was one important exception, which can be detected in the very first pogrom in Elisavetgrad. In the midst of the riot, the authorities were called to the main synagogue, which was surrounded by an angry crowd. The Jews, it was claimed, had been firing at the crowd from the windows of the synagogue with revolvers. The crowd, which presumably was filled with *pogromshchiki*, clearly expected the authorities to take some sort of action. This was a recurrent theme of the pogroms: they turned murderous, not when Jews defended themselves in what may be termed the 'proper' way, but when they used firearms, which was considered as not 'playing by the rules'. Almost all fatalities occurred when the Jews used, or were accused of using, firearms to defend themselves in the course of a pogrom, usually by firing wildly into a mob.

The events at the Elisavetgrad synagogue raise another important point. The pogroms of 1881–2 were directed against Jews, but not Judaism. With the exception of an unconfirmed episode, crowds broke the windows of synagogues but did not enter them or vandalize their contents. (This was no longer the case in the pogroms of 1903 and thereafter.) In particular, Torah scrolls were not touched. Had the pogroms been directed against the Jews primarily as a religious community, we might have expected greater disrespect to be shown to Jewish ritual items. The few episodes of vandalizing of *eruvim* in the Kingdom of Poland do not disprove this thesis, as I have argued above: they were directed against symbols of Jewish 'sovereignty'. In the main, the failure to attack Jewish religious symbols in 1881–2, even when this would have entailed no risk to the *pogromshchiki*, is a telling point which argues against the claim that the pogroms were motivated by some form of 'traditional Russian Orthodox religious Antisemitism'.[31]

An examination of the authors of the pogroms should provide a warning against the over-intellectualizing of the motives of the *pogromshchiki*. The participants of rural pogroms are easily dealt with. They were peasants, often in transit from one place to another. The attacks were commonly directed against victims other than 'their' Jews, such as the local tavern-keeper. Attacks were triggered by the examples of pogroms in larger urban centres. Underlying many episodes were incredible rumours about a tsarist *ukaz* to 'beat the Jews'. Such rumours, which promised the peasants such desiderata as land or freedom from taxes from the 'Good Tsar', were a recurrent phenomenon of peasant life and lore.[32] Those involving the pogroms had a special flavour, often tied to questions of status and respect. There were not only claims that the pogroms were in retaliation for the Jews having killed the 'Orthodox Tsar', but also for having insulted the Tsar in some vague way. The Tsar was inviting his faithful subjects to put the Jews in their place.[33]

The predominant element in all urban pogroms were the so-called 'barefoot brigade', comprising the town proletariat, vagrants, migrant workers, demobilized soldiers and other unsettled elements. Despite the most active search of the authorities, outside agitators and instigators were never found. The urban intelligentsia were rarely involved. On the other hand, 'polite society' turned out in large numbers to watch the pogroms, reinforcing the idea that they were seen as a form of holiday entertainment.[34] All contemporary descriptions of the pogroms depict them as anarchistic revels, rather than ideological protests. For most of the participants, it seems, the pogroms were a form of carnival, of role-reversal, of 'the world turned upside down'.

Conclusions

The purpose of the above survey has not been to deny the violence and suffering endured by the Jews of the Russian Empire during the pogrom waves of 1881–2. Still less does it seek to prove that the Jews were some-how to blame, directly or indirectly, for these events. It does attempt to deny that pogroms can be attributed to 'the tradition of Russian religious Antisemitism'. While a religious element is not lacking in the formulation of the Jew as an outsider and 'the Other' in the popular outlook, other factors must be given greater consideration. The pogroms were not anomalous events, but had analogues in other episodes of interethnic violence and social rivalries within the Russian Empire. The pogroms were not instigated. They lacked a strong ideological under-pinning. Rather, they were an outbreak of popular violence which always lay close to the surface of Russian life. The time of the first out-break – Easter – was one closely connected with riot and carnival and alcoholic excess. If religion played a role in triggering pogroms, it was religion reinforced with strong spirits. Once pogroms broke out in these conditions, Jews were an obvious target. They were an outsider group, restricted by the law of the land, who nonetheless managed to gain domination over Christians, as they were able to remind their non-Jewish neighbours. The pogroms provided an opportunity to put the Jews in their place by attacking real and symbolic elements of their 'domin-ation', the tavern and the market-stall, as well as destroying their ill-gotten gains. Educated observers interpreted this as 'the people responding to Jewish exploitation', but it might better be seen as popular awareness of a rare opportunity by the lower orders to 'put the Jews in their place'.

In short, rather than an episode of religious fanaticism, or a link in an age-old pattern of physical violence of Christians against Jews, the pogroms of 1881–2 should be placed within the complex context of interethnic relations and popular violence in an Empire which was undergoing rapid and disorienting social and political change.

Notes

*Research for this article was assisted by grants from the British Academy, the US National Endowment for the Humanities, the Leverhulme Trust, and funding from the Dean's Fund, the Graduate School Travel Fund, and the Institute for Jewish Studies at University College London.

1. J. Klier, 'Traditional Russian Religious Antisemitism', *The Jewish Quarterly*, **46** (2), (Summer 1999) 29–34.
2. There was a tradition of anti-Jewish violence in Poland and the Ukraine, but not when they were under Russian administration and control. For a very suggestive treatment of the events of 1648, see J. Raba, *Between Remembrance and Denial: The Fate of the Jews in the Wars of the Polish Commonwealth during the Mid-Seventeenth Century as Shown in Contemporary Writings and Historical Research* (New York, 1995). There were isolated attacks upon Jews in Odessa (1821, 1848, 1859, 1871) and Akkerman (1865). The Odessa events are explored in my chapter 'The Pogrom Paradigm in Russian History', in J. D. Klier and S. Lambroza (eds), *Pogroms: Anti-Jewish Violence in Modern Russian History* (Cambridge, 1991) pp. 13–42.
3. In official parlance, however, the term *bezporiadki* ('disorders', always used in the plural) was invariably employed. This term was official jargon for any outbreak of mass violence.
4. I. Michael Aronson has effectively repudiated the possibility of a pogrom conspiracy in his book *Troubled Waters: The Origins of the 1881 Anti-Jewish Pogroms in Russia* (Pittsburgh, PA, 1990).
5. I plan to publish shortly a study of the events of 1881–2, based on archival evidence, which will document this thesis.
6. All dates, unless otherwise indicated, are based on the Julian calendar then in use in Russia, which in the nineteenth century was 13 days behind the Gregorian calendar of the west.
7. See J. D. Klier, 'The Russian Press and the Anti-Jewish Pogroms of 1881', *Canadian–American Slavic Studies*, **17** (1) (1983) 199–221, and Klier, 'The Pogrom Paradigm', 34.
8. The authorities conducted a love–hate relationship with the tavern. Even while recognizing its role as a flashpoint for violence – and ordering that all taverns be closed at the first hint of disorders – they also accepted that alcohol was an essential ingredient of the peasant way of life. In addition, the state liquor monopoly accounted for almost a third of the national budget. See the important study – curiously negligent of the involvement of the Jews – by D. Christian, *Living Water: Vodka and Russian Society on the Eve of Emancipation* (Oxford, 1990).
9. G. Ia. Krasnyi-Admoni, *Materialy dlia istorii antievreiskikh pogromov v Rossii*, vol. 2 (Petrograd-Moscow, 1923) pp. 20, 243. Hereafter, *K-A*.
10. Aronson, *Troubled Waters*, pp. 108–11. Contemporaries of the events were also well aware of this phenomenon.
11. It should be emphasized that, given the scale of the events of 1881–2, and the murder and destruction which would characterize future pogroms, the initial wave had very few fatalities, perhaps 50 persons all told, of whom at least half were rioters shot by the army.

12. See, for example, Omeljan Pritsak, who argues, unconvincingly in my opinion, for central planning and against a significant role of Ukrainians in the urban pogroms, 'The Pogroms of 1881', *Harvard Ukrainian Studies*, **11**, 1–2 (1987) 8–43.

13. *Elizavetgradskii vestnik*, **51**: 13/V/1881.

14. P. A. Zaionchkovskii, *Krizis samoderzhaviia na rubezhe 1870–1880-kh godov* (Moscow, 1964) p. 313.

15. It should be remembered that a large percentage of the Russian army was utilized to maintain domestic order in the Empire. Despite the stereotype of Russia as a 'police state', the local police of the Empire were far too under-manned for the task of law and order, as well as notoriously venal and corrupt. The peasant community preferred its own system of '*samosud*', or vigilante justice, to the official justice system. Day-to-day, random violence was part of the daily life of all the communities of the Russian Empire, including the Jews. This is what I would term the 'dialogue of violence' within and between communities. For an important study of this theme in medieval context, see D. Nirenberg, *Communities of Violence: Persecution of Minorities in the Middle Ages* (Princeton, NJ, 1996).

16. J. D. Klier, 'Evreiskii vopros v slavianofil'skoi presse, 1862–1886 gg.', *Vestnik evreiskogo universiteta v Moskve*, **1** (17) (1998) 49–50.

17. Upon conversion to any Christian faith, an individual ceased to be a 'Jew' in the eyes of the Russian law. See J. D. Klier, 'The Concept of "Jewish Emancipation" in a Russian Context', in O. Crisp and L. Edmondson (eds), *Civil Rights in Imperial Russia* (Oxford, 1989) pp. 121–44.

18. See H. Rogger, *Jewish Policies and Right-Wing Politics in Imperial Russia* (Basingstoke and London, 1986) pp. 113–75.

19. J. D. Klier, *Imperial Russia's Jewish Question* (Cambridge, 1995) pp. 311–20.

20. J. D. Klier, 'What Exactly was a Shtetl?', in G. Estraikh and M. Krutikov (eds), *The Shtetl: Image and Reality* (Oxford, 2000) pp. 23–35.

21. For two examples, see P. Herlihy, *Odessa: A History, 1794–1914* (Cambridge, MA, 1986) and M. F. Hamm, *Kiev: A Portrait, 1800–1917* (Princeton, NJ, 1993). The domination of the marketplace by Jews does not, of course, mean that the Jews as a whole were well-off. The Jewish population of the Empire went through explosive growth in the nineteenth century, rising from approximately one million in 1800 to more than five million in 1900. The Russian economy did not grow rapidly enough to absorb this surplus population. A large percentage of the Jewish population was extremely poor.

22. *Gosudarstvennyi Arkhiv Rossiiskoi Federatsii* (Moscow), 2-oe deloproizvodstvo, fond 102, opis' 38, delo 681, chast' 1 (1881), l. 7. Hereafter *GARF*.

23. One such example comes from Nlowa, in the Kingdom of Poland, where worshippers leaving a church service attacked Jewish market traders. *GARF*. f. 102, op. 39 (1882), d. 280, ch. 1, l. 5ob. Incidents such as this were used to justify the inclusion in the notorious May Laws of 1882 of a ban on Sunday trade by Jews until the end of church services.

24. During the uneasy year of 1882, there were a number of incidents in Poland of attacks on *eruvim*. *GARF*, f. 109, op. 39 (1882), d. 280, ch. 11, l. 1-ob. While such actions might be seen as religiously motivated, I would rather view them as attacks on 'Jewish space' in the village, and an effort to assert non-Jewish control over such space.

25. For example, in the report of the governor of Vitebsk province, contained in *GARF*, f. 542, op. 1, d. 175 (1882), l. 18o.

26. *K-A*, pp. 242–3.

27. *GARF*, f. 109, op. 38 (1881), d. 197, ll. 17–28.

28. See Christian, *Living Water* and 'The Black and Gold Seals: Popular Protests against the Liquor Trade on the Eve of Emancipation', in E. Kingston-Mann and T. Mixter (eds), *Peasant Economy, Culture and Politics of European Russia, 1800–1917* (Princeton, NJ, 1991) pp. 261–93.

29. A. Zel'tser, 'Pogrom v Balte', *Vestnik evreiskogo universiteta v Moskve*, **3**(13) (1996) 40–63.

30. More sophisticated academic studies of the phenomenon of rape were spurred by S. Brownmiller's, *Against Our Will: Men, Women and Rape* (New York, 1975). For a discussion of the significance of Brownmiller's work, and some of her critics, see R. Porter, 'Rape – Does it have an Historical Meaning?', in S. Tomaselli and R. Porter (eds), *Rape – An Historical and Social Enquiry* (Oxford, 1986) pp. 216–36.

31. One can debate whether the stoning of synagogue windows, which also occurred in 1881–2, should be viewed as an attack upon a Jewish religious symbol or included in the category of assaults on 'Jewish space'. See an episode in Poltava province, *GARF*, f. 102, op. 39, d. 280, ch. 12 (19 May 1882), l. 7.

32. See D. Field, *Rebels in the Name of the Tsar* (Boston, MA, 1976) for a discussion of the theme of 'naïve monarchism'.

33. See *GARF*, f. 102, op. 38, d. 680 (1881), l. 33–35ob; ibid., d. 681, ch. 1 l.9-ob.

34. Ibid., l. 16ob.

Part II
Round Table Discussion

9

Religious Violence in Past and Future Perspective

Christopher Andrew

Large-scale violence was one of the defining characteristics of the twentieth century. In addition to witnessing the two most widespread and destructive conflicts in world history, the century also saw the emergence of states which used violence as an instrument of policy on an unprecedented scale. Communist one-party states alone were directly responsible for the death of almost one hundred million of their own citizens.[1] Despite some large-scale outbreaks of religious violence (as, for example, during the partition of India in 1947), only a small proportion of the worst twentieth-century violence was motivated by religion. The most homicidal rulers of the century – Adolf Hitler, Joseph Stalin, Mao Zedong and Pol Pot – wished to wipe out religious practice (save, in Hitler's case, where it could be adapted to the glorification of the Nazi regime).

Despite their hostility to religion, however, the tyrannical one-party states of the twentieth century provide some insight into the roots of religious violence. All claimed both to represent absolute truth and to have the right to impose it on others. Religions, like states, are most prone to violence when they make such absolutist claims. In imposing Stalinism on the states of the Soviet Bloc after the Second World War, Stalin was implementing – probably without realizing it – the sixteenth-century principle, *cuius regio eius religio* ('as country so religion'). The only effective antidote to both religious and political violence is the contrary principle of toleration. Without the emergence of religious toleration in early modern Europe, multi-party democracy would have been impossible. Nineteenth-century British parliamentary democracy was built on Catholic and Jewish emancipation as well as on the extension of the franchise.

The victory of religious toleration over religious fanaticism, however, remains far from complete. During the final two decades of the twentieth century there was an unexpected resurgence of religious violence, which began with Ayatollah Khomeini's Shi'ite revolution in Iran. On 1 April 1979, after an Iranian referendum had produced a massive majority in favour of an Islamic Republic, Khomeini declared that 'the government of God' had begun. His declaration represented a revolution within Shi'ism as well as in Iran; hitherto the mullahs had seen participation in government as demeaning to their spiritual authority. Once in power, religious violence became part of their policies. Even before the beginning of the war with Iraq in September 1980, Khomeini was calling – unsuccessfully – for Iraq's Shi'ite majority to launch a *jihad* against the infidel regime of Saddam Hussein. Infidels and heretics at home – the Baha'is chief among them – were subjected to a reign of terror. During the final months of his life, a decade after the Iranian Revolution, Khomeini was still preaching violence in the name of Islam. In February 1989 he issued a *fatwa* calling for the assassination of Salman Rushdie and all those involved in the publication of *The Satanic Verses* who were 'aware of its content'. Khomeini's last will and testament, published after his death in June 1989, called on his followers to 'sacrifice their own lives and those of their dear ones' in the struggle against 'enemy conspiracies and world-devouring America'.[2]

Despite the declining appeal of religious fanaticism in Iran itself since Khomeini's death, there is no shortage of websites which continue to defend the legitimacy of Khomeini's call to religious violence. According to the website of the 'United Muslims of America':

> Although [the February 1989 *fatwa*'s] demand remained unfulfilled, it demonstrated plainly the consequences that would have to be faced by any aspiring imitator of Rushdie, and thus had an important deterrent effect. Generally overlooked at the time was the grounding of the Imam's *fatwa* in the existing provisions of both Shi'i and Sunni jurisprudence; it was not therefore innovative. What lent the *fatwa* particular significance was rather its issuance by the Imam as a figure of great moral authority.

Khomeini's example helped to inspire a series of fundamentalist (mostly Shi'a) terrorist groups to take up his call to arms against the American 'Great Satan' and its allies. A decade before the Iranian revolution there was not a single religious or cult-based terrorist group anywhere in the world. As recently as 1980 only two of the world's 64 known terrorist groups were religious. Over the next decade and a half, however, Shi'a groups alone were responsible for over a quarter of the deaths from terrorism. Sheik Omar Abdel Rahman, the Egyptian Sunni cleric found guilty of inspiring the 1993 bombing of the World Trade Center in New York, told his followers, 'We have to be terrorists . . . . The Great Allah said, "Make ready your strength to the utmost of your power, including seeds of war, to strike terror into the enemies of Allah" '. Probably the most influential terrorist at the beginning of the twenty-first century is the wealthy Saudi, Osama bin Laden, founder of the 'World Islamic Front for *Jihad* against Jews and Crusaders', who in 1996 declared a holy war against the United States and the West.[3] Though on the FBI's 'Ten Most Wanted' list, and believed to have been involved in a series of terrorist attacks from the 1993 World Trade Center bombing to the attack on USS *Cole* in Yemen in 2000, bin Laden is, according to BBC and other news reports, 'a hero for many young people in the Arab world'.[4]

Though the main support for religious violence in recent years has come from Islamic extremists, it has also found homicidal adherents in other religious traditions. Hindu extremists in India have carried out a series of attacks on Christians and Muslims.[5] In 1995 the Israeli prime minister, Itzhak Rabin, was assassinated by a Jewish extremist, Yigal Amir, who was intent on disrupting the Middle East peace process. Amir told police that he had been following 'orders from God'. Timothy McVeigh, who was chiefly responsible for the bombing of an Oklahama City federal office building, also in 1995, with the loss of 168 lives, was closely linked to the Christian Patriot movement which uses Christian scripture to justify its paranoid call-to-arms. Central to the beliefs of Islamic, Jewish and Christian exponents of religious violence is a hatred of those who think differently, for which they claim divine authority. As 'Meggie' of the US 'Christian Identity' movement explains on the World-Wide Web:

> The Bible is chock full of things we are to hate. . . . Is there a thinking person in this country that would say we have justice anymore? Can they really not say this country is going to hell in a hand basket at breakneck speed? Can they say that Yahweh reigns supreme here? I don't think so. Could our problems stem from our refusing to hate? What does Yahweh say? HATE EVIL.   (Amos 5: 15)[6]

The objects of such hatred, however, are in large part figments of the disturbed imaginations of the fanatics who preach religious violence. Both 'Meggie' and bin Laden view the world through the distorting prism of their own warped conspiracy theories. To justify their homicidal anger, they need to convince themselves that they are waging a holy war against a vast conspiracy of evil. Conspiracy theory, like hatred, is an essential constituent in the mind-set of the advocates and practitioners of religious violence.[7] Antisemitism too has been fuelled by conspiracy theories, from the medieval 'blood libel' (which, as Geoffrey Alderman shows below, surfaced in Pontypridd as late as 1903)[8] to the mythical *Protocols of the Elders of Zion* which appear to have influenced Adolf Hitler.

Contemporary religious and cult-based terrorism is a much more serious menace than the secular terrorist groups who captured the headlines only a generation ago. For most of the twentieth century, terrorist groups (unlike one-party states) rarely sought to kill more than handfuls of people. Their main aims were usually to cause panic and publicize their causes rather than to bring about major massacres. Those who believe they are doing the will of God and combating the forces of Satan, however, are not content simply with sowing terror. The former Hezbollah leader, Hussein Massawi, summed up his aims by saying, 'We are not fighting so that the enemy offers us something. We are fighting to wipe out the enemy.' According to Yoshihiro Inouye of the homicidal Japanese cult, Aum Shinrikyo, 'We regarded the world outside as evil, and destroying the evil as salvation'. Seen in the long-term perspective which has often been lacking in the study of terrorism, there is nothing very surprising about such delusions. The aim of many fanatics in the era of religious warfare was to exterminate, rather than merely to terrorize, their opponents.[9]

Current religious violence, particularly in its most dangerous terrorist manifestations, stems from an ancient tradition of Holy Terror. As Jonathan Riley-Smith argues below, it was not until the sixteenth century that the idea of violence for the common good began to replace violence in the name of God.[10] Though the word 'terrorism' did not yet exist, some of the words we now use to describe terrorism, violence and fanaticism derive from ancient and medieval practitioners of Holy Terror. The first 'assassins' were a radical offshoot of the Shi'a Ismaili sect who between 1090 and 1272 sought to assassinate crusaders. The word 'thug' (appropriately used below by Geoffrey Alderman to refer to the perpetrators of anti-Jewish violence in London's East End) derives from an Indian religious cult, founded in the seventh century, which for the next 1200 years carried out ritual murders as sacrifices to the Hindu goddess Kali. The Zealots were millennarian Jewish terrorists (or freedom fighters) who from AD 66 to 73 waged a campaign of assassination against their Roman rulers.[11] Religious violence is probably as old as religion. Throughout history, as Deirdre Burke argues in this volume, that violence has stemmed not from believers' disregard for their beliefs but from a fanatical defence of 'what they considered to be the truth'.[12]

The closing years of the twentieth century witnessed innumerable conferences summoned to discuss the 'challenges of the twenty-first century'. Most such conferences seem to have paid scant regard to the past experience of the human race, preferring instead to dwell on globalization, information technology and other current preoccupations. At a conference in New York, however, the Holocaust survivor and human rights activist, Elie Wiesel, offered a different vision of the future. 'The principal challenge of the twenty-first century', he said, 'is going to be exactly the same as the principal challenge of the twentieth century: How do we deal with fanaticism armed with power?'[13] Wiesel is surely right. Today's most notorious fanatics seem much less dangerous as individuals than Hitler, Stalin, Mao and Pol Pot. But the power available to them is potentially terrifying. Aum Shinrikyo's use of sarin on the Tokyo underground in 1995 may well be seen by future historians as the moment at which terrorist groups began to move from the old technology of the bomb and the bullet to new weapons of mass destruction.

For those religious extremists who, like Hussein Massawi, seek to 'wipe out the enemy', these new weapons are likely to exercise a fatal attraction. Religious violence in the twenty-first century is highly unlikely to limit itself to the use of the bomb and the bullet.

Notes

1. S. Courtois (ed.), *Le livre noir du communisme* (Paris, 1997), p. 14.
2. On the Khomeini revolution, see, *inter alia*, R. Wright, *In the Name of God: The Khomeini Decade* (London, 1990).
3. B. Hoffmann, 'Responding to Terrorism Across the Political Spectrum', *Terrorism and Political Violence*, **6** (3) (1994); I. O. Lesser et al., *Countering the New Terrorism* (Santa Monica/Washington, 1999).
4. 'Who is Osama bin Laden?', BBC News Online, 20 December 2000.
5. See, for example, the latest Human Rights Watch report on India: www.hrw.org/wr2k/Asia-04.htm
6. Cited by Anthony Lake, *6 Nightmares* (Boston, New York and London, 2000), pp. 22–3.
7. Both were also central to the mind-sets of Hitler, Stalin, Mao and Pol Pot.

8.   See below, p. 181.
9.   C. Andrew, 'The Future of European Security and the Role of Intelligence', *Irish Studies in International Affairs*, **8** (1997) 55–6.
10. See below, p. 184.
11. Bruce Hoffmann, *Inside Terrorism* (London, 1997), ch. 4.
12. See below, p. 188.
13. Cited by R. James Woolsey in H. Shukman (ed.), *Agents for Change: Intelligence Services in the 21st Century* (London, 2000) p. 48.

10
Some Thoughts on Anti-Jewish Violence in Modern Britain

Geoffrey Alderman

The title of this conference refers to the phenomenon of 'religious violence'. I take it that this phrase is meant to refer to violence which is inspired by religious antipathies, or which is interpreted as having been so inspired, even though the perpetrators might have denied that this was the case. Using these rough definitions, I want to draw our attention to instances of anti-Jewish violence in modern Britain – that is, the period following the granting to professing Jews of full civic and political equality in the mid-nineteenth century.

The two hundred or so years which separated Jewish emancipation from the Cromwellian Readmission were punctuated with random acts of violence against Jews; the antisemitism of the streets was (and I am sorry to say still is) a feature of Britain's social landscape. From time to time politicians for their own ends exploited the prejudices to which these acts gave rise. The most notable such instance was the 'Jew Bill' riots of 1753, which have been studied in some depth.[1] But although, in that case, political opponents of the Whig administration whipped up anti-Jewish feeling as a stick with which to beat the government, the prejudices which were let loose did not extend to (for example) serious demands for the Jews to be expelled from Britain, or for such freedoms as the Jews already possessed to be taken from them. The Jews were exploited as scapegoats, and a very modest reform of the law governing naturalization of foreign-born Jews had to be repealed (1754). Thereafter, deprived of its political justification, the violence quickly died away.

During the second half of the eighteenth century Anglo-Jewry prospered. A community of some 8000 souls at the time of the 1753 riots had swollen to perhaps as many as 30,000 by 1815. We read of sporadic violence against Jews and Jewish property during the French revolutionary wars; in 1813 the Birmingham synagogue was attacked. But it is more likely that these acts of aggression originated in xenophobia than in antipathy to the Jews as individuals or the practice of the Jewish religion. Jews enlisted in unprecedented numbers for service in the British armed forces. The road to Jewish emancipation proved a long one. But even those die-hards who opposed the right of professing Jews to sit in Parliament were willing to grant them lesser freedoms. The entire campaign for Jewish emancipation, lasting some 30 years from the granting of Catholic emancipation, passed without any significant incident of anti-Jewish violence. And press reports of anti-Jewish violence in continental Europe – pre-eminently Russia, of course – found widespread condemnation in British society.[2]

Between about 1880 and 1914 a 'native' Anglo-Jewish population of some 60,000 persons was swamped by the immigration to Britain of something like twice that number of mostly poor Jews from central and eastern Europe. What is perhaps more than a little remarkable is that the agitation at the beginning of the twentieth century against unrestricted immigration by aliens to Britain, which culminated in the passage of the Aliens Act in 1905, and which was itself characterized by a strident anti-Jewish discourse, was also noteworthy, however, for the absence of physical violence against Jews. The British Brothers League, established in 1901, both exploited and fostered anti-Jewish prejudice, which fed on widespread popular hostility to the immigration to Britain of Jews from Russia and central Europe over the previous two decades, and on 'rich-Jew' antisemitism which was a feature of left-wing opposition to the Boer War.

Paradoxically, however, the twentieth century has witnessed more anti-Jewish violence in Britain than the nineteenth, and perhaps even than the eighteenth. What is more, this violence has to a measurable extent been explicitly grounded in religious bigotry, rather than in xenophobia.

On Saturday 19 August 1911, what the then Home Secretary, Winston Churchill, described as a 'pogrom' broke out against the Jewish communities living in the Western Valleys of Monmouthshire.[3] Jewish shops were attacked in Tredegar over that weekend, and the rioting spread to Ebbw Vale, Rhymney and other valley towns early the following week, spilling over into Glamorgan before the week was out. These riots have attracted a great deal of scholarly attention.[4] They provide the only example, this century, of racial or ethnic violence in the United Kingdom judged to be of such severity that the military had to be summoned to quell them, because the police were unable to do so.

They arose from a variety of causes, not least of which were the economic and social tensions, and consequent severe industrial unrest, that were endemic in the Welsh coalfields in the years immediately preceding the Great War. I do not wish here to re-examine these riots and their multiple contexts, but rather to stress the religious motifs which underpinned the riots themselves and their historical background. I do so the more urgently because, at the time, their specifically 'Jewish dimension' was either minimized or denied altogether (by both Jews and non-Jews), and because, much more recently, an attempt has been made by a revisionist scholar to reinforce the legitimacy of these denials.[5] In my view, although social and economic tensions were the immediate triggers of the riots, anti-Jewish prejudice had already been made semi-respectable, in Wales, through the activities of Christian groups, most notably the Baptists. In 1868 a Cardiff Baptist minister and his wife were alleged to have abducted for the purposes of conversion a Jewish girl, whose parents subsequently brought (and ultimately lost) a lawsuit. In September 1903 the *Jewish Chronicle* carried a report of a garbled version of the blood libel at Pontypridd.[6] The following year there was an attempt at Llanelli to ban *shechita*, the Jewish religious method of animal slaughter. As to the 1911 riots themselves, although the property of non-Jews as well as Jews became targets of the rioters, it is clear that Jewish property was the prime target. At Tredegar only Jewish shops were attacked; at Ebbw Vale 'the cry of the mob . . . was one long denunciation of Jews'.[7] The riots commenced, at Tredegar, with the singing of 'several Welsh hymn tunes'.[8] There were threats that the synagogue at Ebbw Vale would be burnt.[9] That there were acts of kindness by Christians towards the Jewish victims of the riots is beyond doubt. Many Christians, including Baptists, were horrified by the riots, and especially by their anti-Jewish aspect. But not all. When the Monmouthshire Welsh Baptist Association, meeting at Blackwood, near Bargoed, on 6 September, was asked to pass a resolution expressing sympathy with the Jews, several ministers of religion and others took exception to the motion; one delegate argued that 'Resolutions did more harm than good, and they encouraged the Jews. There were about 100 Jews at Tredegar now, and if they had many more resolutions they would have 500 there.' The resolution was allowed to drop.[10]

I want to turn from these dramatic events to the phenomenon of fascism in England in the 1930s, and to briefly mention one aspect of the rise of the British Union of Fascists (BUF) which it seems to me has been underexplored, and which is worthy of further investigation. I refer to the endorsement which the BUF had from certain ministers of

Notes

1. See, especially, T. W. Perry, *Public Opinion, Propaganda and Politics in Eighteenth-Century England: A Study of the Jew Bill of 1753* (Cambridge, MA, 1962), and G. Alderman, *The Jewish Community in British Politics* (Oxford, 1983) pp. 5–7.
2. W. D. and H. L. Rubinstein, 'Philosemitism in Britain and in the English-speaking World, 1840–1939', *Jewish Journal of Sociology*, **40** (1998) 11–15.
3. Public Record Office, Home Office Papers, HO45/10656/212470 (251–380), file 317 of 1911: minute by Churchill, 29 August 1911.
4. G. Alderman, 'The Anti-Jewish Riots of August 1911 in South Wales', *Welsh History Review*, **6** (1972) 190–200. See also C. Holmes, 'The Tredegar Riots of 1911: Anti-Jewish Disturbances in South Wales', *Welsh History Review*, **11** (1982) 214–25; and A. Glaser, 'The Tredegar Riots of August 1911', in U. R. Q. Henriques, *The Jews of South Wales: Historical Studies* (Cardiff, 1993), pp. 151–76.
5. W. D. Rubinstein, 'The Anti-Jewish Riots of 1911 in South Wales: a Reexamination', *Welsh History Review*, **18** (1997) 667–99.
6. According to the blood libel, Jews used the blood of Christians for ritual purposes.
7. *Jewish Chronicle*, 1 September 1911, 1; *Daily News*, 21 August 1911, 1.
8. *Jewish World*, 25 August 1911, 9; *Jewish Chronicle*, 25 August 1911, 8.
9. *South Wales Argus*, 23 August 1911, 3.
10. *The Times*, 7 September 1911, 6; *Jewish Chronicle*, 8 September, 11.
11. T. P. Linehan, 'The British Union of Fascists in Hackney and Stoke Newington, 1933–1940', in G. Alderman and C. Holmes (eds), *Outsiders and Outcasts: Essays in Honour of William J. Fishman* (London, 1993) p. 144.
12. Ibid., p. 145.
13. *Kilburn Times*, 9 June 1944, 3.

11
Religious Violence
Jonathan Riley-Smith

A consensus among Christians and liberals in western Europe and America on the legitimate use of force, which has prevailed for the last century, has become so embedded in our thinking that we do not realize how recently it developed, how weak it is conceptually, how impractical it has proved to be and how close it is to collapse. It goes as follows. Violence is intrinsically evil, but can be allowable in the last resort when one is confronted by a greater evil. Recourse to violence may therefore be condoned by God – or, to the liberal, may be acceptable – when the consequence of failing to employ it will be a situation worse than the existing status quo. Certain conditions have also to be met. It must have a just cause. The power which authorizes it must be recognized as having the right in international law to do so. The men and women managing and perpetrating it must have right intentions. The premises underlying this consensus could hardly be further from the assumptions of patristic and medieval writers that the use of force could be pleasing to God when it was directed in support of, and in accordance with, the divine plan for mankind, could be perpetrated on God's direct or indirect authority and, being itself morally neutral, drew its ethical colouring from the intention, bad or good, of the perpetrator.

It was only in the sixteenth century that the idea of violence employed in the name of God or Christ began to be replaced by that of violence employed in the name of 'the common good', the defence of which was the prerogative of every community and the justification of which had to be in accordance with accepted earthly laws. A catalyst appears to have been the atrocities committed against the Indians in the New World and the critical reaction to them of the Spanish Dominican Francisco de Vitoria, who developed notions to be found in embryonic forms in the work of Thomas Aquinas. With Vitoria's followers, particularly Francisco Suarez and Baltasar Ayala, 'just war' arguments moved from the field of moral theology to that of law, to be elaborated within decades by Alberico Gentili and Hugo Grotius.[1]

As for the view that violence is intrinsically evil, this does not even seem to have been held by enlightenment thinkers. Its transference from pacifism, accompanied by the argument that violence can nevertheless be condoned as the lesser of evils, was, I believe, an achievement of the peace movement which swept Europe and America after the Napoleonic Wars and split into two wings, pacifist and moderate, in the 1830s.

The just war theory most of us take for granted is, therefore, a very recent introduction. It is conceptually weak because it is hard to conceive of anything being intrinsically evil, given that evil has always been held to be negative, being nothing more than the absence of good in the same way as darkness is the absence of light or cold the absence of heat. Scripture is far more ambivalent on the subject than most realize; and although Christians are obviously bound to adapt moral laws sensibly and with charity to their own special situations, it is surely dangerous to base a moral generalization on the contradiction that God can condone what to him is an evil.

The application to nineteenth-century just war theory, moreover, of the criteria of just cause, legitimate authority and right intention, which were originally conceived, it should be remembered, in the context of a much more positive approach to force, raised within decades practical problems which proved themselves to be intractable. The precondition of a just cause imagined an unlikely world in which independent parties of equal worth in the sight of God coexisted and in which an 'injurer' could be recognized as transgressing norms of behaviour before the use of defensive force by the 'injured' could be justified. It has often been pointed out that the just cause involves the unlikely presupposition of unqualified justice on one side and unqualified injustice on the other. As early as the 1860s international lawyers were abandoning discussion of its merits and were beginning to concentrate on getting international agreement on rules of war that might ameliorate the suffering that accompanied conflict, rather than on basic principles. The criterion of legitimate authority proved its fallibility over and over again as it was exploited by rogue states and was stretched to its limits in attempts to justify 'righteous rebellion'. As for right intention, no consensus could be reached in the debates on proportionality with respect to nuclear deterrence. The extraordinarily confused justifications given for the Gulf War by some leading churchmen demonstrated, if demonstration was needed, that modern just war theory was going nowhere.

So it is not surprising that a shift in thought now seems to be taking place, in a period when interconfessional violence seems to be spreading, whether in acts of international terrorism or civil wars or religious uprisings or the suppression of heresy – in Lebanon, Northern Ireland, the Balkans, Indonesia, Chechnya, Egypt, Sudan, Nigeria, Israel, Iraq and Afghanistan. At this stage it is often hard to tell how influenced the feelings being expressed in these conflicts are by nationalism, but the prospect of major wars of religion, while still distant, must be greater than at any time in the last two centuries. Alongside the resurgence of ideas of *jihad* in Islam, Christ as an authorizer of violence has returned to the scene in the writings of those associated with the militant Christian liberation movement in the 1960s and early 1970s, leading some moral theologians at the time, who were reluctant to abandon the notion that force was intrinsically evil and claimed to be writing 'mature ethics', to suggest that on this issue God could order men to sin.

There has also emerged a train of thought in which subjective ethical judgements, manifesting themselves in the notion of crimes against humanity and in judgements by international tribunals, have been incorporated into international law. This seems to have set in motion a process which has led again to the waging of ethical wars proclaimed by international bodies. So in the place of crusades proclaimed by the papacy, warfare justified by humanitarian aims, the definition of which rest on a moral consensus, has reappeared, authorized by the Security Council of the United Nations or NATO.

If I am right and just war theory is again transforming itself, it is exceptionally important for theologians, philosophers, historians and scholars of politics and international studies to keep their heads and to face up to the issues as objectively as they can, because no one else will do so. A first step might be for the religious among us to discard the sentimental and unhistorical assumption that the faiths we represent are unambiguously pacific ones. A second would be to shun emotional breast-beating or romance and to study carefully the many cases of interconfessional war and persecution in the past in order to analyse what engendered them. A third would be to disseminate the results of our research as widely as possible in order to encourage open debate and to demonstrate how quickly xenophobia can spread and the ease with which communities can be deceived by it. Only by recognizing the potentiality for violence which seems to lie at the heart of most ideologies is there any hope of a way forward.

Note

1. H. Duchhardt, 'La Guerre et le droit des gens dans l'Europe du XVIe au XVIIIe siècles', in P. Contamine (ed.), *Guerre et concurrence entre les états européens du XIVe au XVIIIe siècles* (Paris, 1998) pp. 340–56; B. Hamilton, *Political Thought in Sixteenth-century Spain: A Study of the Political Ideas of Vitoria, De Soto, Suarez and Molina* (Oxford, 1963) *passim*.

12

Religious Violence: Educational Perspectives

Deirdre Burke

Violence in the name of religion

The prophet Isaiah was warned that the people would: 'Hear, indeed but do not understand; See, indeed, but do not grasp' (Isaiah 6: 9). These words can help us to understand how throughout history it has been possible for human beings to perceive the divine message to require them to act in ways that we may now consider to be against the will of God. Religious violence throughout history has not been caused by the 'moral incontinence' of believers failing to live up to what they perceive to be good. It has, rather, been occasioned by believers being true to what they considered to be the truth, and we are astounded by the lengths to which believers went to defend their faith from both internal and external attack. Untold numbers have given up the comforts of hearth and home to travel over land and sea to support the cause of their faith and engage in acts of violence against a perceived enemy. This phenomenon is not just one which affected our ancestors but is one which continues to blight the world in which we live.

Religious scholars today respect the scholarship of German theologians from earlier decades of this century in subjecting the Bible and theological topics to rigorous scrutiny, but we are challenged by the failure of many of these scholars to recognize that the Nazi policy towards Jews was wrong. Indeed we must share Fisher's question to understand how German society, which was 95 per cent Christian at that time,[1] could actively or passively support laws which discriminated against Jews and finally led to the Final Solution: 'What was missing (or, more chillingly present) in the Christian education they had received for centuries that allowed them to remain blind to what they were doing? Or indifferent to what others were doing in their name?'[2]

It is certain that many of the episodes that are covered in this conference would have been justified by religious authorities at the time
they occurred. Such episodes were understood within the framework of
divine revelation, and may even have been required by religious doctrines. The thought that scholars would meet centuries hence to analyse
critically their actions would have been inconceivable to the instigators.

Today we are operating from very different perspectives. We are heirs
to a philosophical tradition which questions the relationship between
religion and morality, and many have expressed 'profound dissatisfac-
tion with some particular religious teaching or practice'.[3]

Lewis demonstrated that religious teaching could be morally harmful,
stating that 'The institutional life of religion has also sheltered, and
sometimes actively encouraged, grievous practices of which persecution
is an outstanding example.'[4] Thus, we are involved in the calling to
account of religious authorities for actions of violence which were carried
out in the name of religion.

A shift in thinking

The discussion on religious violence and the terms we are using show
a radical shift in our thinking about which actions should be termed
violent. The terms that we are using show this change in perspective –
the shift from self-proclaimed 'holy war' to religious violence in itself
speaks volumes. 'Violence' is a negative term that has the connotation
of illegal or inappropriate use of force: 'the illegal exercise of physical
force, an act of intimidation by the show or threat of force'.[5]

Drawing from the language developed to study the Holocaust,[6] we are
likely to apply Hilberg's terms 'perpetrator', 'victim' and 'bystander' to
the study of acts of violence.[7] In this scenario religions which provided
the leadership, motivations and rewards are seen in the negative role of
the 'perpetrators', holding responsibility for acts of violence. Those who
have been portrayed in negative terms and were on the receiving end
of the acts of violence are now regarded sympathetically and termed
'victims'. In addition, those who were not actively involved in violent
incidents may be subject to criticism as 'bystanders'. Thus, contemporary scholarship in one field can be seen to influence other areas
and raises new questions about the way the past is to be perceived
and studied.

We are also influenced by the pluralistic nature of society. Western
societies recognize the legitimate existence of a range of religions; in
England and Wales the Education Reform Act of 1988 required new

Agreed Syllabuses for Religious Education to cover Christianity and 'other principal religions'. In the past only one religion was regarded as the truth and there was a commitment to eradicate falsehood, both within the true faith and outside. Developments in Christian theology show this movement: John Hick's 'Copernican revolution in theology' makes it possible for Christians to move from a Christocentric to a theocentric world-view. This takes the believer from a world-view that accepts only its own view of God as the truth to one which recognizes that 'God has many names'.[8]

An additional factor in raising awareness about religious violence is the excess of blood that has been shed in the twentieth century. Lemkin applied the term 'genocide' in 1944 to refer to the attempt to eradicate a specified group.[9] The term is used to refer to killings of people (men, women and children) in the name of ideology, race, ethnicity or religion. This phenomenon casts its shadow over us – and leads us to question the role of religion in contemporary ethnic cleansing (with its quasi-religious trappings).

In light of these, and many other, factors we are now asking questions about such events/incidents which the people at the time and many followers since have regarded as appropriate if not good. We are now inclined to ask: 'Is it ever right for violent actions to be carried out in the name of religion if they cause suffering or death?'

The role of education

The first strand concerns research within faiths and the education of believers. Post-Holocaust theological reflection has led Jews and Christians to a re-examination of their faiths. Rubenstein questioned how the Jewish self-definition of 'chosen' contributes to negative Christian views of Jews. He asked: 'Does the way Jews regard themselves religiously contribute to the terrible process?'[10] A positive answer led him to re-evaluate the concept of 'chosen' to eliminate elements of self-belief which led to explosive reactions. Within Christianity attempts have been made by many theologians to identify the origins of the 'teaching of contempt' in scripture and tradition. Littell challenged Christians to reassess their faith and 'to understand the Jewish experience'.[11]

These theological developments are mirrored by liturgical responses within the faiths. In Progressive Judaism, for example, attempts have been made to reinterpret passages, that could on the surface lead to violence, so that the Torah truly becomes an 'Etz Hayim', tree of life, to fulfil the words in the liturgy: 'It is a tree of life to those who hold it fast, and all who cling to it find happiness. Its ways are ways of pleasantness and all its paths are peace'.[12]

The second strand of education concerns schooling. Watson argued that Religious Education can make a 'positive contribution to world peace'. She stated: 'The power of religion is frighteningly dynamic, and it has often been called the most dangerous force in the world.'[13] In order to harness this power for good it is necessary to educate to remove ignorance. She claimed: 'Education in the West is at a critical point... Universities as well as schools are criticised for failing to expose students to values and turning out "knowledgeable barbarians", who have no real understanding of their society.'[14] The opportunity to link Religious Education with citizenship is now present in the school curriculum in England and Wales. Pupils have the opportunity to 'learn from' religions and apply key religious principles to society.

The following extracts, from a study conducted with 14-year-old pupils following a visit to *Anne Frank: A History for Today* exhibition within Religious Education, show how pupils were able to identify key 'citizenship' issues in relation to racism, indifference and individual responsibility.

> We need to be aware of what happened, and how it was supposed to be good, but was actually very bad. People need to know that prejudice and discrimination only lead to anger, hurt and maybe war again.

> I still find it difficult to understand how anyone could be so cruel, and why the rest of the world were so helpless.

> To always speak out against something which we don't believe to be right and not just 'turn the other cheek' we must never let it happen again.[15]

Any exploration of 'religious violence' should have the aim of learning from the mistakes of the past, so that the cognitive failures of our ancestors do not return to haunt us. Education, both within religious communities and in schools, is essential to ensure that individuals and communities are able to develop the skills to discriminate between good and evil.

Notes

1. J. S. Conway, *Nazi Persecution of the Churches 1933–1945* (New York, 1968).
2. E. J. Fisher, 'Why Teach the Holocaust?', paper at *Peace/Shalom after Atrocity* (Seton Hill Conference: Yad Vashem Pedagogical Centre, 1989) 2.
3. H. D. Lewis, *Philosophy of Religion* (London, 1965) p. 256.
4. Ibid., p. 256.
5. B. Kirkpatrick (ed.), *Cassell Concise English Dictionary* (1992) p. 1468.
6. The term Holocaust will be used in this paper, as the intended frame of reference encompasses all victims of Nazi persecution; if the intention had been to focus on Jewish experiences and perspectives, the term Sho'ah would have been more appropriate.
7. R. Hilberg, *The Destruction of the European Jews* (New York, 1979).
8. J. Hick, *God has Many Names: Britain's New Religious Pluralism* (London, 1980).
9. G. S. Yacoubian, 'Underestimating the Magnitude of International Crime: Implications of Genocidal Behavior for the Discipline of Criminology', *Injustice Studies*, 1 (1997) [on-line journal].
10. R. Rubenstein, *After Auschwitz* (London, 1992) p. 13.
11. F. H. Littell, *The Crucifixion of the Jews* (Macon, 1986).
12. *Siddur Lev Hadash*, Union of Liberal and Progressive Synagogues (London, 1995) p. 489.
13. B. Watson, *Education and Belief* (Oxford, 1987) p. 104.
14. Ibid., p. 1.
15. D. M. Burke, 'The Holocaust in Education: Teacher and Learner Perspectives', University of Wolverhampton PhD thesis (1998) p. 247.

Bibliography

Selection of English Translations of Sources and Further Secondary Reading

See also full annotations accompanying each chapter.

Primary sources

Abelard, Peter, *Dialogue of a Philosopher with a Jew and a Christian*, trans. P. J. Payer, Medieval Sources in Translation, 20 (Toronto, 1979); see also the rendering of the text as *Collationes*, ed. and trans. J. Marenbon and G. Orlandi, Oxford Medieval Texts (Oxford, 2001).

Aelred of Rievaulx, *On Jesus at the Age of Twelve*, trans. Th. Berkeley, in *Treatises: The Pastoral Prayer*, Cistercian Fathers Series, 2 (Spencer, MA, 1971) pp. 3–39.

Anselm of Canterbury, *The Prayers and Meditations of St Anselm with the Proslogion*, trans. B. Ward (Harmondsworth, 1973).

Bernard of Clairvaux, *On Loving God*, ed. and trans. E. G. Gardner (London, [1916]); *On Loving God with an Analytical Commentary* by E. Stiegman, Cistercian Fathers Series, 13B (Kalamazoo, MI, 1995).

Benton, J. F. (trans.), *Self and Society in Medieval France: The Memoirs of Abbot Guibert of Nogent* (New York, 1970).

Berger, D. (ed. and trans.), *The Jewish–Christian Debate in the High Middle Ages: A Critical Edition of the Nizzahon Vetus* (Philadelphia, PA, 1979).

Chazan, R. (ed.), *Church, State and Jew in the Middle Ages* (New York, 1980).

Edwards, J. (trans. and ed.), *The Jews in Western Europe, 1400–1500* (Manchester, 1994).

Eidelberg, S. (ed. and trans.), *The Jews and the Crusaders: The Hebrew Chronicles of the First and Second Crusades* (Madison, WI, 1977).

Maccoby, H. (ed. and trans.), *Judaism on Trial: Jewish–Christian Disputations in the Middle Ages* (Rutherford, NJ, 1982).

Resnick, I. M. (trans.), *On Original Sin and a Disputation with the Jew, Leo, Concerning the Advent of Christ, the Son of God: Two Theological Treatises of Odo of Tournai* (Philadelphia, PA, 1994).

William of St Thierry, *Enigma of Faith*, trans. J. D. Anderson, Cistercian Fathers Series, 9 (Washington, DC, 1974).

——, *The Golden Epistle: A Letter to the Brethren at Mont Dieu*, trans. Th. Berkeley, Cistercian Fathers Series, 12 (Spencer, MA, 1971).

——, *Mirror of Faith*, trans. T. X. Davis, Cistercian Fathers Series, 15 (Kalamazoo, MI, 1979).

Secondary Sources

Abulafia, A. Sapir, *Christians and Jews in Dispute: Disputational Literature and the Rise of Anti-Judaism in the West (c. 1000–1150)* (Aldershot, 1998).

——, *Christians and Jews in the Twelfth-Century Renaissance* (London, 1995).

Abulafia, D., *Spain and 1492*, Headstart History Papers (Bangor, 1992).

Brundage, J., *Medieval Canon Law and the Crusader* (Madison, WI, 1969).

Chazan, R., *Barcelona and Beyond: The Disputations of 1263 and its Aftermath* (Berkeley and Los Angeles, CA, 1992).

——, *Daggers of Faith: Thirteenth-Century Christian Missionizing and Jewish Response* (Berkeley and Los Angeles, CA, 1989).

——, *European Jewry and the First Crusade* (Berkeley and Los Angeles, CA, 1987).

——, *God, Humanity, and History: The Hebrew First Crusade Narratives* (Berkeley and Los Angeles, CA, 2000).

——, *Medieval Stereotypes and Modern Antisemitism* (Berkeley and Los Angeles, CA, 1997).

Cohen, J., *The Friars and the Jews: The Evolution of Medieval Anti-Judaism* (Ithaca, NY, 1982).

——, *Living Letters of the Law: Ideas of the Jew in Medieval Christianity* (Berkeley and Los Angeles, CA, 1999).

Cohen, M. R., *Under Crescent and Cross: The Jews in the Middle Ages* (Princeton, NJ, 1994).

Cole, P., *The Preaching of the Crusades to the Holy Land, 1095–1270* (Cambridge, MA, 1991).

Constable, G., *The Reformation of the Twelfth Century* (Cambridge, 1996).

Dobson, R. B., *The Jews of Medieval York and the Massacre of March 1190*, Borthwick Papers, no. 45 (York, 1974).

Erdmann, C., *The Origin of the Idea of Crusade*, trans. M. W. Baldwin and W. Goffart (Princeton, NJ, 1977).

Jacobs, L., *The Jewish Religion: A Companion* (Oxford, 1995).

Jordan, W. C., *The French Monarchy and the Jews: From Philip Augustus to the Last Capetians* (Philadelphia, PA, 1989).

Katz, J., *Exclusiveness and Tolerance: Jewish–Gentile Relations in Medieval and Modern Times* ([London], 1961).

Klier, J. D., *Imperial Russia's Jewish Question, 1855–1881* (Cambridge, 1995).

——, *Russia Gathers her Jews: The Origins of the 'Jewish Question' in Russia, 1772–1825* (DeKalb, IL, 1986).

Klier, J. D. and Lambroza, S. (eds), *Pogrom: Anti-Jewish Violence in Modern Russian History* (Cambridge, 1992).

Langmuir, G., *History, Religion and Antisemitism* (Berkeley and Los Angeles, CA, 1990).

——, *Toward a Definition of Antisemitism* (Berkeley and Los Angeles, CA, 1990).

Maier, C. T., *Crusade Propaganda and Ideology: Model Sermons for the Preaching of the Cross* (Cambridge, 2000).

Marcus, I. G., *Rituals of Childhood: Jewish Acculturation in Medieval Europe* (New Haven, CT, 1996).

McCulloh, J. M., 'Jewish Ritual Murder: William of Norwich, Thomas of Monmouth, and the Early Dissemination of the Myth', *Speculum*, **72** (1997) 698–740.

Nirenberg, D., *Communities of Violence: Persecution of Minorities in the Middle Ages* (Princeton, NJ, 1996).

Riley-Smith, J., *The First Crusade and the Idea of Crusading* (London, 1986).

——, 'The First Crusade and the Persecution of the Jews', in W. J. Sheils (ed.), *Persecution and Tolerance*, Studies in Church History, vol. 21 (Oxford, 1984) pp. 51–72.

Rubin, M., *Gentile Tales: The Narrative Assault on Late Medieval Jews* (New Haven, CT, 1999).

Stow, K. R., *Alienated Minority: The Jews of Medieval Latin Europe* (Cambridge, MA, 1992).

Tolan, J., *Petrus Alfonsi and his Medieval Readers* (Gainesville, FL, 1993).

Trachtenberg, J., *The Devil and the Jews: The Medieval Conception of the Jew and its Relation to Modern Antisemitism* (New Haven, CT, 1943).

Trautner-Kromann, H., *Shield and Sword: Jewish Polemics against Christianity and the Christians in France and Spains, 1100–1500* (Tübingen, 1993).

Yuval, I. J., 'Easter and Passover as Early Jewish–Christian Dialogue' and 'Passover in the Middle Ages', in P. F. Bradshaw and L. A. Hoffman (eds), *Passover and Easter: Origin and History to Modern Times – Two Liturgical Traditions*, vol. 5 (Notre Dame, IN, 1999) pp. 98–124 and 127–60.

——, 'Jews and Christians in the Middle Ages: Shared Myths, Common Language', in R. S. Wistrich (ed.), *Demonizing the Other: Antisemitism, Racism and Xenophobia* (Singapore, 1999) pp. 88–107.

Index